12th International Conference on the Health Effects of Incorporated Radionuclides (HEIR 2018)

BIO Web of Conferences Volume 14 (2019)

Fontenay-aux-Roses, France
8 - 11 October 2018

ISBN: 978-1-5108-8618-6

Printed from e-media with permission by:

Curran Associates, Inc.
57 Morehouse Lane
Red Hook, NY 12571

Some format issues inherent in the e-media version may also appear in this print version.

Printed by Curran Associates, Inc. (2019)

For additional information, please contact EDP Sciences – Web of Conferences at the address below.

EDP Sciences – Web of Conferences
17, Avenue du Hoggar
Parc d'Activité de Courtabœuf
BP 112
F-91944 Les Ulis Cedex A
France

Phone: +33 (0) 1 69 18 75 75
Fax: +33 (0) 1 69 28 84 91

contact-edps@webofconferences.org

Additional copies of this publication are available from:

Curran Associates, Inc.
57 Morehouse Lane
Red Hook, NY 12571 USA
Phone: 845-758-0400
Fax: 845-758-2633
Email: curran@proceedings.com
Web: www.proceedings.com

TABLE OF CONTENTS

PREFACE

THE 12 TH INTERNATIONAL CONFERENCE ON HEALTH EFFECTS OF INCORPORATED RADIONUCLIDES

HEIR 2018

The 12th international conference on the health effects of incorporated radionuclides (HEIR 2018), took place in Fontenay-aux-Roses, from October 8 to October 11, 2018. Jointly organized by the French Institute of Radioprotection and Nuclear Safety (IRSN) and the Commissariat à l'Energie Atomique et aux Energies Alternatives (CEA), this international conference was the first one organized in France. The series began in 1974 in Alta (Utah, USA). Further ones followed almost every 3-4 years: Neuherberg (Germany) in 1976, Lisbon (Portugal) in 1977, Lake Geneva (Wisconsin, USA) in 1981, Neuherberg in 1984, Bethesda (Maryland, USA) in 1988, Heidelberg (Germany) in 1994, Tokyo (Japan) in 1999, Neuherberg in 2004, Santa Fe (New Mexico) in 2009 and Berkeley (California, USA) in 2013.

In the 1970s, the first HEIR conference has focused on radium and thorotrast health consequences; since then the spectrum of the conferences has extended to other radionuclides, such as radon and risk of cancer among uranium miners, mixed oxides of uranium and plutonium and their biological effects, more recently radioactive iodine and potential health effects after the Fukushima accident. The wide scope of technical, biological tools (for example 'omics technologies') and specific models developed contribute largely to address health and biological questions due to the human contamination by radionuclides.

The program was divided into 7 sessions: *biokinetics*, *dose assessment*, epidemiology, biological effects, *medical countermeasures*, *nuclear medicine* and *what's next*. For the first time in this series of conferences, a session was dedicated to the medical use of radioisotopes.

More than 140 scientists coming from Algeria, Belgium, Canada, Czech Republic, France, Germany, Hungary, Japan, Kazakhstan, Romania, Russia, Spain, Switzerland, United Arab Emirates, UK and USA, attended the conference. Over almost four days, more than 80 presentations or posters were given and discussed. All these contributions are collected in this new issue of Bio Web of conferences.

We would like to thank the members of the scientific committee for the evaluation process of the abstracts and the building of the scientific program. They are:

Rebecca Abergel, US

Makoto Akashi, Japan

Manuel Bardies, France

Luiz Bertelli, US

Jean-Marc Bertho, France

Eric Blanchardon, France

Xavier Castagnet, France

Augusto Giussani, Germany

Laurence Lebaron-Jacobs, France

Jin Kyung Lee, Korea

Valérie Lemercier, France

Balazs Madas, Hungary

Florence Menetrier, France

Brian Moyer, US

François Paquet, France

Tony Riddell, UK

Thierry Sarrazin, France

Patsy Thompson, Canada

Anne Van Der Meeren, France

Richard Wakeford, UK.

Moreover, we warmly thank the members of the organizing committee: Céline Dinocourt, Sarah Le Hir, Caroline Salhab, Fabrice Ecrabet, Keltoum Mediana for providing a very nice welcome, organizing secretary, partnership, communication and media.

Professional Societies SFRP, SoFRa, SFMN and SFPM are thanked for their scientific or technical support, so as Cevidra Society for its grant. We would like to thank specifically the Board of the Multidisciplinary European Low-Dose Initiative MELODI for its financial support having allowed the Young Investigator Award.

Finally, we hope that you will read with pleasure these proceedings of HEIR 2018.

Jean-Marc Bertho and Florence Menetrier
Co-chair of the HEIR2018 conference

The new ICRP biokinetic and dosimetric models

François Paquet[1]*
[1]*IRSN, PSE ENV-SRTE BP 3, 13115 Saint Paul Lez Durance Cedex, France*
**The author thanks all the members of the Internal Dose Coefficients Task Group of ICRP, who are the collective authors of this work*

1. The use of the models

In internal dosimetry, no operational dose quantities have been defined which directly provide an assessment of equivalent dose or effective dose. Different methods are therefore applied to assess the doses due to radionuclides in the human body. They are mostly based on various activity measurements and the application of biokinetic and dosimetric models.

Biokinetic models that are developed for individual elements and their radioisotopes are used to calculate the total number of transformations occurring within specific tissues, organs or body regions (source regions) during a given period of time (usually 50 years for adults, or to age 70 years for children) by determining the time-integrated activity in each source region. Dosimetric models, based on Male and Female Reference computational phantoms and Monte Carlo radiation transport codes, are used to calculate the deposition of energy in all important organs/tissues (targets) for transformations occurring in each source region, taking account of the energies and yields of all emissions. At this stage, sex-specific absorbed doses in each target organ or tissue resulting from a nuclear disintegration in each source region are calculated. The radiation weighting factors are then applied to determine sex-specific committed equivalent doses to an organ or tissue, and the tissue weighting factors are finally applied to determine the sex-averaged committed effective dose.

2. New biokinetic and dosimetric models developed by ICRP

ICRP has been very active for many decades in providing models for the calculation of internal doses. The Human Respiratory Tract Model (HRTM) and the Human Alimentary Tract Model (HATM) were described in ICRP Publication 66 and ICRP Publication 100 [1-2]. Systemic models for individual elements were described in several publications providing also dose coefficients for workers and the members of the public for intakes of radionuclides by inhalation and ingestion (See section 3) [3-10]. Finally, the ICRP Publication 88 and 95 [11-12] described the biokinetic data for the transfer of radionuclides to milk and for the transfer of radionuclides to the embryo, fetus and newborn child. Since these publications, ICRP has continuously improved the accuracy of the models published. New models are being produced and will be used in the new series of ICRP publications (OIR Series and EIR Series). These new models are briefly described below:

Human Respiratory Tract Model

The Human Respiratory Tract Model (HRTM) described in Publication 66 [1] has been updated in ICRP Publication 130 [13] to take account of data accumulated since its publication, although the basic features of the model remain unchanged. Inhaled particles containing radionuclides deposit in the ET airways (nose, larynx, etc.), the bronchial and bronchiolar airways of the lung and the AI region, with deposition in the different regions being mainly dependent on particle size [1]. Removal from the respiratory tract occurs mainly by dissolution and absorption to blood and the competing process of transport of particles to the throat followed by their entry into the alimentary tract. The proportions absorbed to blood or cleared by particle transport depend on the speciation and the solubility of the material, and on the radioactive half-life of the radionuclide. The ICRP model for the respiratory tract is also applied to gases and vapours and to inhalation of radon and its radioactive progeny.

For absorption to blood, the main changes introduced in ICRP Publication 130 are:

- Redefinition of the Type F, M and S absorption defaults: larger rapid dissolution fraction (f_r) values for M and S of 0.2 and 0.01, rather than 0.1 and 0.001, respectively, with lower rapid dissolution rate (s_r) values of 3 d^{-1} for M and S, and 30 d^{-1} for F, rather than 100 d^{-1};
- Material-specific parameter values for f_r, s_r and the slow dissolution rate (s_s) in selected cases where sufficient information is available (e.g. forms of uranium);
- Element-specific values of s_r and the bound state parameters, f_b and s_b, where sufficient information is available; and
- Revised treatment of gases and vapours in which solubility and reactivity are defined in terms of the proportion deposited in the respiratory tract. The default assumption is 100% deposition (20% ET_2, 10% BB, 20% bb and 50% AI), and Type F absorption. The SR-0, -1, -2 classification described in *Publication 66* has not been found to be helpful and is not used.

For clearance by particle transport, the main changes are:

- More realistic clearance from the nasal passage, including transfer from the anterior to the posterior region, based on recent human experimental studies;
- Revised characteristics of slow particle clearance from the bronchial tree based on recent human experimental studies. It is now assumed that it occurs only in the bronchioles rather than as a particle size dependent phenomenon throughout the bronchial tree; and
- Longer retention in the AI region of the lung, with a revised model structure, based on recent data including long-term follow-up of workers exposed to insoluble 60Co particles, and plutonium dioxide.

Human Alimentary Tract Model

The ICRP Publication 30 [3] model of the Gastrointestinal tract has been replaced by the HATM described in ICRP Publication 100 [2]. The main features of the HATM can be summarised as follows:

- Inclusion of all alimentary tract regions: oral cavity, oesophagus, stomach, small intestine, right colon, left colon and rectosigmoid (the sigmoid colon and rectum);
- A default assumption that absorption of an element and its radioisotopes to blood occurs exclusively in the small intestine, i.e. the total fractional absorption, f_A equals the fractional absorption from the small intestine, f_{SI}. A model structure that allows for absorption in other regions, where information is available;
- A model structure that allows for retention in the mucosal tissues of the walls of alimentary tract regions, and on teeth, where information is available; and
- Explicit specification of the location of target regions for cancer induction within each alimentary tract region.
- ICRP Publication 100 [2] gave preliminary values of electron and alpha particle absorbed fractions for stem cell layers in alimentary tract regions. ICRP Publication 133 [14] provided new calculations for both particle types and for both content and wall sources. For regions within the small intestine, new models of segment folding were also implemented.

Systemic models

A systemic model describes the time-dependent distribution and retention of a radionuclide in the body after absorption to blood and systemic circulation, and its excretion from the body. In contrast to ICRP's current and past biokinetic models describing the behaviour of radionuclides in the respiratory and alimentary tracts, ICRP's systemic models have generally been element-specific with regard to model structure as well as parameter values. A single generic model structure that depicts all potentially important systemic repositories and paths of transfer of all elements of interest in radiation protection would be too complex to be of practical use. However, generic model structures have been used in previous ICRP documents to represent the systemic biokinetics of groups of elements, typically chemical families, known (or expected to have) qualitatively similar behaviour in the body. For example, ICRP Publication 20 [15] introduced a generic model formulation for the alkaline earth elements calcium, strontium, barium and radium, but provided element-specific values for most model parameters. In Parts 1-3 of ICRP Publication 30 [3-5], a model developed for plutonium, including parameter values as well as model structure, was applied to most actinide elements. The use of generic systemic model structures was increased in ICRP reports on doses to members of the public from intake of radionuclides [7-9] and is further expanded in recent ICRP Publications because it facilitates the development, description and application of systemic biokinetic models. An important development is that, as the availability of data allows, models have been made to be more physiologically realistic with regard to the dynamics of organ retention and excretion so that they can more reliably be applied to the interpretation of bioassay data as well as the calculation of dose coefficients.

3. From the models to the dose coefficients

Biokinetic and dosimetric models provide every data needed to calculate the dose from incorporation of the radionuclides. However, because of the complexity of the overall procedure when calculating internal dose, ICRP provided also sets of dose per intake coefficients that allow a direct estimate of the internal dose from knowledge of intake into the body. In addition, for occupational intakes, data are provided to allow intake and dose to be calculated from bioassay measurements. The current work of ICRP is to provide a new set of models and dose coefficients, to take into account the latest ICRP recommendations [16] and up-to-date knowledge in biology, physiology and dosimetry. They are going to be published in the Occupational Intake of Radionuclides (OIR) series and the Environmental Intakes of Radionuclides (EIR) series.

The OIR series replace the ICRP Publication 30 series [3-6] and ICRP Publication 68 [17] to provide revised dose coefficients for occupational intakes of radionuclides (OIR) by inhalation and ingestion. The revised dose coefficients have been calculated using the new biokinetic models described in section 2. The first of five reports in the OIR series, *Publication 130* (ICRP, 2015), includes chapters on control of occupational exposures, biokinetic and dosimetric models, monitoring methods, monitoring programmes and retrospective dose assessment. Subsequent reports provide data on individual elements and their radioisotopes, including biokinetic data and models, dose coefficients and data for bioassay interpretation. OIR Part 2 and Part 3 have been issued as Publication 134 and 137 [18-19], and OIR 4 is nearing completion. In these reports, each element section provides: dose coefficients (committed effective dose and committed equivalent doses to organs or tissues per Bq intake (Sv Bq^{-1}) for inhalation and ingestion of all relevant radioisotopes; dose per content functions [committed effective dose per predicted activity content in the body or in a given organ or per daily excretion (Sv Bq^{-1})]; reference bioassay functions [values of activity (Bq) retained in the body or specific organs, or excreted in urine or faeces, at various times after unit intake (i.e. 1 Bq) by inhalation or ingestion].

The EIR series will replace the ICRP Publication 56 series [7-10] to provide revised age-dependent dose coefficients for members of the public for environmental intakes of radionuclides by inhalation and ingestion. As for the OIR series, revisions have been made to use the updated HATM et HRTM and new systemic models, and considering the age-dependence of biokinetics. The first report in the EIR series will provide an introduction to the report series and include data on individual elements and their radioisotopes, including biokinetic data and models, and dose coefficients. As for the OIR series, each element section will provide dose coefficients: committed effective dose and committed equivalent doses to organs or tissues per Bq intake (Sv Bq^{-1}) for inhalation and ingestion of all relevant radioisotopes.

References

1. ICRP. Human Respiratory Tract Model for Radiological Protection. ICRP Publication 66, Ann. ICRP **24,** 1-3 (1994a)
2. ICRP. Human Alimentary Tract Model for Radiological Protection. ICRP Publication 100. Ann. ICRP, **36,** (1-2) (2006)
3. ICRP. Limits for intakes of radionuclides by workers. ICRP Publication 30, Part 1. Ann. ICRP, **2** (3/4) (1979)
4. ICRP. Limits for intakes of radionuclides by workers. ICRP Publication 30, part 2. Ann. ICRP, **4**, 3/4 (1980)
5. ICRP. Limits for intakes of radionuclides by workers. ICRP Publication 30, Part 3. Ann. ICRP, **6**, 2/3 (1981)
6. ICRP. Limits for intakes of radionuclides by workers. ICRP Publication 30, Part 4. Ann. ICRP **19**,4 (1988)
7. ICRP. Age-dependent dose to members of the public from intake of radionuclides: part 2: ingestion dose coefficients. ICRP Publication 67. Ann. ICRP **23**, 3-4. (1993)
8. ICRP. Age-dependent Doses to Members of the Public from Intake of Radionuclides: Part 3: Ingestion Dose Coefficients. ICRP Publication 69. Ann. ICRP, **25**,1 (1995a)
9. ICRP. Age-dependent Doses to Members of the Public from Intake of Radionuclides: Part 4: Inhalation Dose Coefficients. ICRP Publication 71. Ann. ICRP, **25**,3/4, (1995c)
10. ICRP. Age-dependent Doses to Members of the Public from Intake of Radionuclides: Part 5 Compilation of Ingestion and Inhalation Dose Coefficients. ICRP Publication 72. Ann. ICRP, **26**,1 (1996)
11. ICRP. Doses to the Embryo and Fetus from Intakes of Radionuclides by the Mother. ICRP Publication 88. Ann. ICRP **31,** 1-3 (2001)
12. ICRP. Doses to Infants from Ingestion of Radionuclides in Mothers' Milk. ICRP Publication 95. Ann. ICRP **34**,3/4, (2004)
13. ICRP. Occupational Intakes of Radionuclides, Part 1. ICRP Publication 130. Ann. ICRP **44** (2) (2015)
14. ICRP. The ICRP Computational Framework for Internal Dose Assessment for Reference Adults: Specific Absorbed Fractions. ICRP Publication 133. Ann. ICRP **45**, 2, (2016a)
15. ICRP. Alkaline Earth Metabolism in Adult Man. ICRP Publication 20. Ann. ICRP. (1973)
16. ICRP. The 2007 Recommendations of the International Commission on Radiological Protection. ICRP Publication 103. Ann. ICRP **37**, 2–4, (2007)
17. ICRP. Dose Coefficients for Intake of Radionuclides by Workers. ICRP Publication 68. Ann. ICRP, **24**, 4, (1994b)
18. ICRP. Occupational Intakes of Radionuclides: Part 2. ICRP Publication 134. Ann. ICRP **45**, 3/4, (2016b)
19. ICRP. Occupational Intakes of Radionuclides: Part 3. ICRP Publication 137. Ann. ICRP. **46**, 3/4, (2017)

OBT and human tritium dosimetry

Anca Melintescu[1,*]

[1] "Horia Hulubei" National Institute of Physics and Nuclear Engineering, 30 Reactorului St., POB MG-6, Bucharest-Magurele, RO-077125, Romania

Tritium (^3H) is a radioactive isotope of hydrogen that is ubiquitous in environmental and biological systems. Following the debate regarding the human health risk from exposure to tritium, there have been claims that the former bio kinetic model recommended by the International Commission on Radiological Protection (ICRP) may underestimate tritium doses [1]. Recently, ICRP published its new bio kinetic model for tritium [2]. In a country that develops Canada Deuterium Uranium (CANDU) reactors for production of nuclear energy, tritium is the 'national radionuclide'. Tritium can enter into the human body in many forms, but only tritiated water (HTO) and organically bound tritium (OBT) generate exposures of concern for members of the public. Based on a default OBT intake fraction of 0.5, the new ICRP model [2] increases the tritium dose to workers by 21 % comparing to the former model [1] and that could start a debate. There are some uncertainties associated with OBT metabolism as described by the ICRP's OBT model [2], particularly the retention of OBT in children and the effect of organ growth regarding the OBT retention [3]. The relative depositions of OBT in various organs and tissues are not fully known for all age groups and consequently, there is no age-dependent bio kinetic model for dietary intakes of OBT that accounts for age-related physiological and anatomical variations. HTO turnover in the human body is well understood, but the experimental data for tritium turnover in organic compounds are few, particularly for the long-term bio kinetic components.

Different physiological models regarding the OBT bio kinetics were proposed [4, 5]. A quite recent human bio kinetic model [6] considers the age and gender dependence, as well as the OBT deposition in various organs. The model was favourably tested with both HTO and OBT experimental data in urine for humans following the HTO intakes. The model was also favourably tested with OBT data in urine for humans following the OBT intake. The HTO and OBT data in urine were provided by five volunteers who ate food containing fish from Cardiff Bay, and HTO and OBT in their urine were analysed for up to 150 days [7]. After a short presentation of the model, the model tests with HTO and OBT experimental data following the OBT intake is discussed in relation with the experimental errors due to the measurement techniques. The model also predicts the dynamics of HTO and OBT in urine and blood (red blood cells and plasma) and potentially offers a new approach for bioassay. The need for a standardised bioassay technique is discussed.

References

1. ICRP 1989. Age-dependent doses to members of the public from intakes of radionuclides: Part 1. International Commission on Radiological Protection ICRP Publication **56**. Ann. ICRP 20 (2) (1989)
2. ICRP 2016. Occupational Intakes of Radionuclides: Part 2. ICRP Publication **134**. Ann. ICRP 45(3/4) (2016).

[*] Corresponding author: ancameli@nipne.ro

3. CNSC 2010 Tritium Studies. Health effects, dosimetry and radiological protection of tritium. INFO-0799. Canadian Nuclear Safety Commission (2010), available at http://www.nuclearsafety.gc.ca//pubs_catalogue/uploads/CNSC_Health_Effects_Eng-web.pdf
4. R.B. Richardson, D.W. Dunford, Health Phys. **85** (2003)
5. A. Melintescu, D. Galeriu, H. Takeda Radiat. Prot. Dosim. **127** (2007)
6. D. Galeriu, A. Melintescu, J. Radiol. Prot. **30** (2010)
7. J. Hunt, T. Bailey, A. Reese , J. Radiol. Prot. **29** (2009)

Americium systemic model for rats

Dunstana Melo[1] , *Guthrie* Miller[2], *John* Klumpp[3], *Deepesh* Poudel[3], *Waylon Weber*[4], *Jasen* Swanson[5], and *Raymond* Guilmette[6]

[1]Melohill Technology, LLC, Rockville, MD
[2]Santa Fe, NM
[3]Radiation Protection Division, Los Alamos National Laboratory, Los Alamos, NM
[4] Lovelace Biomedical, Albuquerque, NM
[5]Department of Nuclear Engineering & Health Physics, Idaho State University, Pocatello, ID
[6]Ray Guilmette and Associates, LLC, Perry, ME 04667, USA

1 Introduction

Americium-241 is an internal and external radiation hazard often encountered at nuclear reactors and nuclear weapons facilities. It is a man-made metal usually produced by the decay of ^{241}Pu. In this work, a baseline compartmental model of the distribution and distribution of americium in rat following a systemic intake is derived. We take a novel pharmacokinetic (PK) front-end modeling approach in which physiological knowledge of vascular flow to the tissues and volumes of extracellular fluid (ECF) associated with the tissues are taken into account. The movement of material into ECF spaces is rapid and is completely characterized by vascular flow rates to and from a tissue and the volumes of the tissue and the ECF associated with the tissue [1]. Other transport, presumably involving cellular processes, is modeled in a purely empirical manner. The PK model incorporates the amount of ECF available in each tissue, which the authors anticipate being useful for chelation modelling. It is also shown to provide a significantly better fit to the data when compared to a more traditional biokinetic model.

2 Materials and Methods

The systemic biokinetic model for americium presented here is based on data of distribution and retention of ^{241}Am obtained from studies conducted at Lovelace Respiratory Research Institute (LRRI). Some data of the ^{241}Am distribution in rats have already been published elsewhere [2]; the data are in very good agreement with other publications [3-5]. The data used in the current analysis is the most comprehensive set of data available for ^{241}Am intravenously injected in rats.

The empirical modeling approach is described in Ref. [6]. The model for the animal is broken up into subsystems that interact through a central compartment like plasma or blood, which is fit to the blood data to determine a prescribed function of time plasma(t) or blood(t). Independently for each subsystem, tissues are modeled as sums of a small number of compartments, which interact with the central compartment. The number of compartments used is the minimum needed to obtain a satisfactory value of ChiSq/nData (approximately 1) for the subsystem. This is an imprecise concept that involves some judgement. The Plasma(t) [or Blood(t)] method (sometimes referred to as the open-loop method) allows breaking up an unmanageably large model (on the order of 50 unknown parameters) into manageable pieces with simple enough structure to be amenable to trial-and-error fitting. Furthermore, this method can be iterated, where the functional form of Plasma(t) or Blood(t) is that obtained by solving the entire model self consistently and then reapplied to individual subsystems. The iterated method allows more accurate Markov Chain Monte Carlo (MCMC) modeling of the smaller-dimensionality subsystems.

3 Results

The transfer rates and their uncertainties and correlations were studied using Markov Chain Monte Carlo. Uncertainty is found to be a sensitive function of complexity of the model structure.

Regarding the gross distribution of americium, at 1-hour post IV intake, material has almost entirely left the plasma and relocated to liver (56% in liver) and bone (20% in bone). In the PK modeling approach, material goes rapidly to ECF and then transfers to liver (15 min time scale) and skeleton (30 min time scale). The plasma circulation time (from plasma volume divided by cardiac output) is about 16 seconds or 2×10^{-4} day, and in the PK model the transfer out of blood is on this time scale. In the PK model this early-time transfer is from blood into ECF, and then later, on a 15-30 min time scale, transfer to liver and bone.

Both the non-PK and PK modelling approaches provide a reasonable description of the dataset, but the PK approach has a significantly larger chain-average likelihood. Either model could be used when modelling the perturbation of americium biokinetics by chelation therapy, but the PK model might be useful in predicting the relative chelation in intra-cellular and extra-cellular fluid spaces.

References

1. Miller G, Klumpp JA, Melo D, Poudel D. Health phys. 113(6), 519-526
2. Doyle-Eisele M, Weber W, Melo DR, Guilmette RA. International journal of radiation biology. 1;90(11):1068-74. (2014)
3. Durbin, Patricia W. DISTRIBUTION OF TH2 TRANSURANIC ELEMENTS IN MAMMALS. (1962)
4. Taylor DM. Health phys. 8(6):673-7. (1962)
5. Lloyd RD, Mays CW, Taylor GN, Atherton DR. MJ RETENTION AND DOSIMETRY. (1970)
6. Miller G, Bertelli L, Klare K, Weber W, Doyle-Eisele M, Guilmette R. Health phys. 103(4):484-94. (2012)

Influence of the physico-chemical form on americium biodistribution after wound contamination: re-analysis of experimental data using STATBIODIS

Stephanie Lamart[1,*], *Anne* Van der Meeren[1], *Sophie* Dubois[2], *Solveig* Bibard[1], *Ana Paula* Serond[3], *Jaime* Angulo[1], and *Nina* M Griffiths[1]

[1]Laboratoire de Radio Toxicologie, CEA, Université Paris-Saclay, 91297 Arpajon, France
[2]Anastats, 37220 Rilly sur Vienne, France
[3]Orano, 92400 Courbevoie, France

1 Context

Aging of nuclear fuel and decommissioning of nuclear power plants may give rise of wound contaminations with complex mixtures of actinides. To understand better the actinide behaviour in these conditions and to evaluate the efficacy of decorporation protocols, a rodent model of wound contamination has been developed in the laboratory using different wound types, actinides, and physico-chemical forms [1]. More recently, a statistical tool, described in [2] and hereinafter named STATBIODIS, has been developed to facilitate the analysis of biodistribution data from internal contamination experiments. Hence, STATBIODIS allows re-analyses of data from experiments using different contamination conditions and/or carried out at different time periods.

The present study aimed at evaluating the influence of the physico-chemical form and of the presence of plutonium (Pu) in the contaminant, on the biodistribution of americium (Am) after wound contamination.

2 Material and methods

Retention and excretion data were collected from several series of wound experiments using Am alone or in presence of Pu, and in two physico-chemical forms (nitrate or MOX: mixed oxide of uranium and Pu isotopes, also containing Am-241 as decay product of Pu-241). Results of activity measurements in 189 tissues and 161 urine samples were organized in a standardized table format, which built up to 350 lines of data. Descriptive statistics analysis was carried out for each experimental group using STATBIODIS. Am tissue retention and urinary excretion of each animal were analysed as percentage of Am administered activity. To compare Am relative systemic biodistribution between the selected experimental conditions, skeleton to liver activity ratios were also computed. Statistical tests were used to evaluate the influence of the physico-chemical form and of the

* Corresponding author: stephanie.lamart@cea.fr

presence of plutonium in the contaminant on Am tissue retention and urinary excretion. Non parametric tests essentially adapted to paired data, such as the Wilcoxon test, were applied. Significance level of 0.05 was chosen for bilateral tests and 0.025 for unilateral tests. The false discovery rate correction was applied for multiple comparisons.

3 Results

The 24 h urinary excretion of Am was significantly greater for Am nitrate than for Am from MOX from day 1 to 7. In addition, excretion data were fitted with sums of exponential terms, which showed a greater excretion rate for Am nitrate. Kidney, skeleton and liver retentions were greater for Am nitrate than for Am from MOX at day 1 and also at day 7 after contamination. Skeleton to liver activity ratios were not significantly different between Am nitrate and Am from MOX seven days after contamination. Hence, Am relative biodistribution between skeleton and liver was independent of the initial physico-chemical form, whereas there was overall a greater retention in systemic tissues for Am nitrate than for Am from MOX.

One day after wound contamination, no difference was found between Am nitrate and Am nitrate mixed with Pu in terms of urinary excretion, tissue retention (blood, lymph node, kidney, liver) and skeleton to liver ratio. Overall, the presence of Pu did not modify Am biodistribution after contamination with Am nitrate mixed with Pu after the first 24 hours.

4 Conclusion

In this work, experimental data were re-used to evaluate the influence of the physico-chemical form and of the presence of Pu on the Am biodistribution after wound contamination. Although not all contamination conditions were available across collected data, this re-analysis of biodistribution data using STATBIODIS provided answers to new questions without carrying out any further animal experiments, which is in line with the concept of the 3Rs (Reduce, Refine, Replace) of ethical animal research.

Results presented here were obtained in the frame of the Orano-CEA collaboration agreement.

References

1. N.M. Griffiths, J.C. Wilk, M.C. Abram, D. Renault, Q. Chau, N. Helfer, C. Guichet, and A. Van der Meeren, Health Phys. **103**(2): p. 187-94 (2012).
2. S. Lamart, N.M. Griffiths, N. Tchitchek, J.F. Angulo, and A. Van der Meeren, J Radiol Prot. **37**(1): p. 296-308 (2017).

A method for tracking a case under chelation using urinary excretion measurements

Luiz Bertelli[1,*], *Deepesh* Poudel[1], *John* Klumpp[1], and *Tom* Waters[1]

[1]Los Alamos National Laboratory, Radiation Protection, MS G761, Los Alamos - NM, 87545 USA

1 Introduction

A considerable effort has been made to generate methods and biokinetic models for interpreting the urinary excretion of plutonium under the influence of DTPA [1-9]. The approaches have evolved from empirical [1, 2] to mechanistic [3] and most recently leaning towards pharmacokinetic approaches associating biokinetic compartments from the latest accepted plutonium systemic models [10] with those having more influence on the enhancement of the urinary excretion [4, 5, 6, 9]. In addition, work has been published with practical applications for estimating the influence of the excretion enhancement factor on internal dose estimates [7].

Several methods mentioned above focus on estimates of intake, dose, and efficacy of the chelation therapy only after all urine samples have been collected and/or the effect of chelation has subsided (typically taken to be 100 days). The need for early provisional estimates soon after the administration of DTPA has been proved urgent. The estimates can be refined as more bioassay samples are collected and analysed. The method described below has been developed for intake of plutonium by wounds, but it can be extended to other routes of intake and other radionuclides as well.

2 The proposed method

We propose a mechanistic method for tracking the intake and dose estimates and the efficacy of chelation therapy comprising single or multiple treatments, using the available information from Los Alamos National Laboratory (LANL) cases and a simplified wound model [8, 11, 12].

The wound simulation uses a numerical ordinary differential equations solver [13] applied to the latest plutonium systemic model [10], with one unit of activity initially deposited in the wound. A table containing the predicted activities in urine and in the whole body for the unperturbed case (without chelation) is generated and stored. As the chelation treatments occur, the enhanced excreted activity is calculated against the corresponding values from the unperturbed case, accounting for activities already removed in previous treatments. It uses a pre-established average enhancement factor and a two-term exponential function, like that proposed in the Hall model [2], with coefficients for rapid and slow removal of the chelate and short- and long-term retention half-lives of the chelate

* Corresponding author : lbertelli@lanl.gov

from LANL cases. These parameters will be refined as more bioassay data is collected. The solver is then stopped and the additional activity to be added to the enhanced excretion is selectively removed from the compartments that show the greatest availabilities for removal, i.e., higher transfer rates back to blood and higher activity values. This process is dominated by the amounts in blood, urinary bladder, renal tubules, and rapid-turnover soft tissue. The solver is then restarted, taking into account the delayed effects of each treatment for each subsequent day. Activities in compartments and the corresponding number of nuclear transformations are calculated for later comparison against the results from the unperturbed cases to account for the averted dose.

As real bioassay results become available, a maximum likelihood method is used to fit the intake using the available samples. In this way, preliminary estimates can be made before sufficient bioassay results are available to produce a final, more precise, estimate.

3 Further studies

We intend to study excretion patterns from related cases to assess additional input parameters for the method. We also intend to apply this method to inhalation cases and to other radionuclides using, whenever possible, available information from similar cases to assess input parameters.

References

1. J.J. Jech, B.V. Andersen, K.R. Heid, Health Phys **22** (1972)
2. R.M. Hall, G.A. Poda, R.R. Fleming, J.A. Smith, Health Phys **34** (1978)
3. T.R. LaBone, *Practical applications of internal dosimetry*, (Medical Physics Publishing, Madison, WI, 2002)
4. A.C. James, L.B. Sasser, D.B. Stuit, S.E. Glover, E.H. Carbaugh, Radiat Prot. Dosim. 127 (1–4) (2007)
5. B. Breustedt, E. Blanchardon, P. Berard, P. Fritsch, A. Giussani, M.A. Lopez, A. Luciani, D. Nosske, J. Piechowski, J. Schimmelpfeng, A.L. Serandour, Health Phys **99**, (2010)
6. K. Konzen, R.Brey, Health Phys **108,** 6, (2015).
7. E. Davesne, E. Blanchardon, B. Peleau, P. Correze, S. Bohand, D. Franck, Health Phys. **110**, 6 (2016)
8. D. Poudel, L. Bertelli, J. A. Klumpp, T. L. Waters, Health Phys. **113**, 1, (2017)
9. S. Dumit, M. Avtandilashvili, S.Y. Tolmachev, Health Phys. **115**, 1, (2018)
10. R. W. Leggett, K. F. Eckerman, V. F. Khokhryakov, K. G. Suslova, M. P. Krahenbuhl, and S.C. Miller, Radiat. Res. **164**, 2 (2005)
11. L. Bertelli, T.L. Waters, G. Miller, M.S. Gadd, M. C. Eaton, R.A. Guilmette, Health Phys. **99**, 4, (2010)
12. D. Poudel, J. A. Klumpp, T. L. Waters, L. Bertelli, Radiat. Prot. Dosim. **178**, 2 (2018)
13. A.C. Hindmarsh, Odepack, A systematized collection of ODE solvers, *Scientific Computing*, (North-Holland, Amsterdam, 1983)

Mobility, solubility and biokinetics of ^{238}Pu

John Klumpp[1,*]

[1]Radiation Protection Services, Los Alamos National Laboratory, Mail Stop G761, Los Alamos, NM 87544, United States of America

1 Introduction

Plutonium-238 is commonly used in radioisotope thermoelectric generators for space missions beyond Jupiter, where the sunlight is too weak to use as a power source. The workers who prepare the plutonium heat sources are at risk for intakes of the material, and require routine monitoring. Although chemically identical, experience has shown that the high specific activity of ^{238}Pu dramatically increases its mobility through a laboratory, and alters its solubility both in the laboratory and in vivo. This effect on solubility, in which recoil nuclei gradually increase the solubility of a particle, is particularly important for the biokinetics of ^{238}Pu oxides, which initially exhibit very low solubility in the respiratory tract.

2 Mobility in the laboratory

While the mobility of ^{238}Pu is well known to those who work with it, many of the details are limited to personal correspondence and anecdotes. At the Savannah River National Laboratory, personnel who worked with ^{238}Pu described it as "lifelike," stating that fines appear to "walk" down pipes and "fly" through the air. It was also reported that surfaces were difficult to decontaminate because particles were easily re-suspended into the air and previously decontaminated surfaces were later found to be re-contaminated after hours or days. [1] The United States Department of Energy warms that "microspheres have been observed to climb walls of glass beakers and spread throughout the glovebox" and that, because of internal heating, 100-200 micron spheres can melt through glove material. [2] Correspondence with Los Alamos workers describe instances of ^{238}Pu hot particles dislodging themselves from wipes or tape. In one instance, after decontaminating a wall, the contamination was found to have migrated to a different wall (presumably driven by electrostatic forces). Quantitative studies have shown that ^{238}Pu particles trapped in HEPA filters are re suspended back into the laboratory at a rate of 5% per month by weight, compared to 0.02% per month for ^{239}Pu. [3] Particles can also be transported through HEPA filters by aggregate alpha recoil at a rate dependent on specific activity and number of atoms per aggregate. [4]

3 Solubility and fragmentation

* Corresponding author: jaklumpp@lanl.gov

When a nucleus decays, the recoil nucleus from an alpha decay can damage a particle and increase its solubility. This is particularly important in plutonium oxides, which, when intact, have extremely low solubility. A cone of 50x50 angstroms approximates the volume damaged by a recoiling nucleus of PuO_2. [5]. The ratio of dissolution between $^{238}PuO_2$ and $^{239}PuO_2$ is about 200, consistent with the ratio of decay rates. When aged $^{239}PuO_2$ was dissolved in water, a large amount of dissolution was observed in the first 24 hours, dropping by about 2 order of magnitude the next day. The high initial solubility was attributed to accumulated damage. In $^{239}PuO_2$, the release material was due to damaged regions in the original particle which were subsequently exposed to water [6,7]. In contrast, $^{238}PuO_2$ has been shown to fragment. For example 34-micron particles have been shown to fragment at 0.001% per day. [8]

4 Biokinetics

Unusual biokinetics for plutonium oxides were first observed in 1971 following an exposure to a $^{238}PuO_2$ ceramic. Excretion was below the detectable limit for about 100 days following the incident, and eventually increased to large values, peaking about 3 years following the incident. The observed excretion could be predicted by starting with an initial lung solubility of $1 \times 10^{-6}\ d^{-1}$, increasing gradually to $4 \times 10^{-3}\ d^{-1}$. A mechanistic model was developed for non-ceramic plutonium oxides in beagles [9], and was eventually adapted to fit the urinary excretion observed in cases involved in the 1971 incident [10, 11]. The ICRP66/67 models were calibrated to fit the excretion and autopsy data of a worker who was exposed to ceramic $^{238}PuO_2$, but since this was only based on one individual, this modified ICRP model is not broadly applicable [12]. The latest ICRP publication (in draft) includes dissolution models for ceramic and non-ceramic ^{238}Pu-oxides [ICRP 2018]. However, neither of these is able to predict the observed excretion from a more recent incident at LANL involving several individuals, in which the excretion peaks at about 10 days, then remains constant for many years. We consider various ICRP-compatible dissoloution models which are consistent with all of the observed excretion patterns.

References

1. Congdon JW. SRT-MTS-96-3026. (1996)
2. U.S. Department of Energy. DOE-STD-1128-98. (2003)
3. Icenhour, AS. No. ORNL/TM-2005/22. (2005)
4. McDowell, WJ, FG Seeley, MT Ryan. Health Phys. **32**, 445-447. (1977)
5. Fleischer, R.L. Health Phys. **29**(1), 69–73. (1975)
6. Fleischer, RL, and OG Raabe. Health Phys. 32, 253–257. (1977)
7. Fleischer, RL, and OG Raabe. Health Phys. **35**, 545–548. (1978)
8. Clinard, FW, & Rohr, DL. LA-UR-80-2956; CONF-801124-15. (1980).
9. Mewhinney, JA, & Diel, JH. Health Phys, 45(1), 39-60. (1983)
10. Guilmette RA, Griffith, WC, Hickman, AW. (1994). Rad prot dos. 53(1-4), 127-131.
11. Hickman AW, Griffith WC, Roessler GS, Guilmette RA. Health phys. **68**(3), 359-370. (1995)
12. James, AC, Filipy, RE, Russell, JJ, McInroy, JF. (2003). Health phys, **84**(1), 2-33.
13. ICRP, Occupatioal Intakes of Radionuclides Part 4. (In Preparation)

From *in vivo* to *in vitro* models to assess bioavailability properties of Plutonium compounds

Olivier Grémy[1], *Nina* Griffiths[1], *Laurent* Miccoli[1], and *Anne* Van der Meeren[*,1]

[1]Laboratoire de Radiotoxicologie, CEA, Paris-Saclay University, 91297 Arpajon - FRANCE

The understanding of the biodistribution of actinides following internal contamination is an important step in the process of dose estimation and risk assessment, as well as in determining an appropriate decorporation treatment. Dose calculation is based, in part, on transferability properties of radiocontaminant. For this purpose ICRP classifies materials according to their absorption from lungs into blood in three types: Fast (F), Moderate (M), and Slow (S). Plutonium (Pu) can exist in multiple forms with differing physico-chemical properties in the work place. It is thus beneficial to develop robust in vitro models as a first approach to predict Pu compound behavior in vivo and to facilitate dose calculation and decisions for appropriate therapy.

For an estimation of Pu biodistribution following inhalation, pulmonary administration in rats is the most common approach. However, animal models remain time consuming and are not appropriate for the study of a great number of compounds. In addition, an a priori knowledge for the categories of a Pu compound involved in a contamination incident would be of great help for an early dose assessment. Therefore, we have developed in vitro models to i) mimic uptake and release by macrophages [1] and ii) determine bioavailability properties of radiocontaminants to predict the in vivo behavior [2].

In the present study, using in vivo and in vitro models, we evaluated the behavior of a Pu compound of unknown physico-chemical properties. Pu solution with unknown properties was intra-tracheally administered in rats, or used for in vitro contamination of macrophages as well as in the acellular model for bioavailability prediction (fig 1). Data were then compared to our historic data obtained using three forms of Pu (colloid, nitrate, citrate) of distinct solubility properties [3] and that we chose to represent the three ICRP transportability classes (respectively S, M and F).

Two weeks following pulmonary contamination in rats, retention in lungs, bone and liver was evaluated by alpha activity measurement using liquid scintillation counting. Broncho-alveolar lavages (BAL) were carried out to recover alveolar macrophages and quantification of the retained Pu activity. Both the human macrophage-like THP-1 cells and alveolar macrophages from naïve non-exposed rats were exposed to the unknown Pu

* Corresponding author: anne.vandermeeren@cea.fr

compound in solution for two hours to assess Pu uptake. Thereafter, Pu-loaded macrophages were cultured in a transwell® culture system and Pu release was measured over 7 days. Finally, in a third experimental approach, agarose gels were prepared so Pu compound solution was entrapped in the static phase representing the retention compartment, and the released fraction in the dynamic phase was an indication of the intrinsic solubility properties of the contaminant.

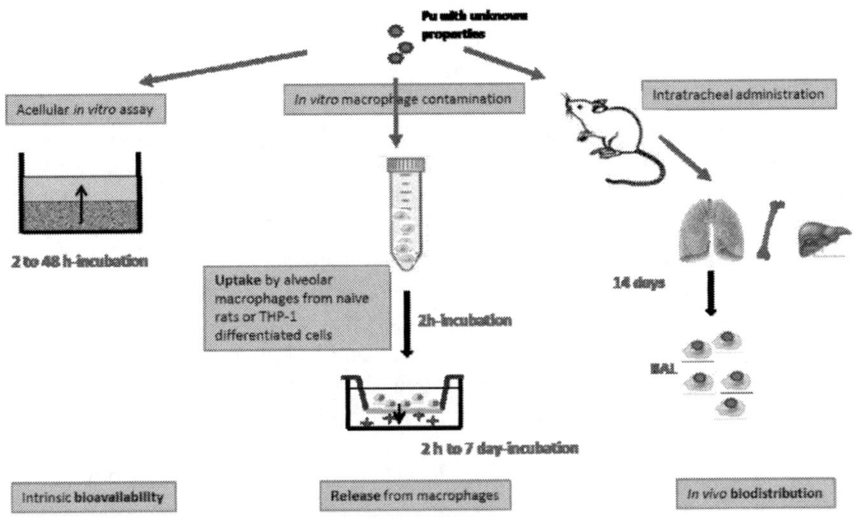

Fig.1. Design of the experimental approach

By comparing the data to previous results obtained in the laboratory using Pu as colloid, nitrate or citrate form, and whatever the model used, we demonstrated that the unknown Pu compound behaved very similarly to Nitrate. Therefore, were able to classify the Pu compound as a type M default absorption type, according to ICRP classification. These results suggest that the in vitro methods could be used as a first approach to predict the biodistribution of a contaminant, and help during the dose assessment process to classify the contaminant according to its solubility properties. In addition, the development and use of in vitro models to predict retention and propose transportability classes fulfill the aims of the 3Rs (Reduce, Refine, Replace) with regard to animal experimentation.

Results presented here were obtained in the frame of the AREVA NC-CEA collaboration.

[1] A. Van der Meeren, A. Moureau, D. Laurent, P. Laroche, JF. Angulo-Mora. Toxicol. in Vitro **37**, 25 (2016)

[2] CEA-AREVA NC International Patent Application WO2017109024 published June 29, 2017.

[3] A. Van der Meeren, O. Grémy, D. Renault, A. Miroux, S. Bruel, N. Griffiths, F. Tourdes. J. Radiat. Res. **53**, 184 (2012)

Biokinetics of soluble plutonium after wound injury treated with Ca-DTPA

Maia Avtandilashvili[1,*] and *Sergei Y.* Tolmachev[1]

[1] U.S. Transuranium and Uranium Registries, College of Pharmacy, Washington State University, 1845 Terminal Drive, Suite 201, Richland 99354, WA, USA

1 Introduction

The mission of the United States Transuranium and Uranium Registries (USTUR) is to study the uptake, translocation, retention and excretion (biokinetics), and tissue dosimetry of uranium, plutonium, americium, and other actinides in occupationally exposed volunteer Registrants (tissue donors). These individuals were mainly exposed to various types of plutonium material with inhalation and wound as primary routes of intake. The USTUR holds records on exposure history and bioassay measurements, as well as post-mortem tissue radiochemical analysis results for 19 individuals who had documented intakes of ^{239}Pu due to contaminated wounds. For 8 individuals, internal deposition resulted from a single wound injury, and three of them were treated with decorporation therapy. In this study, USTUR Case 0303 was used to study biokinetics of soluble Pu after wound intake.

1.1 Case description

The USTUR whole-body donor (Case 0303) was employed at nuclear defence facility for 30 years. While working in a glove-box, he accidentally punctured his finger on a sharp object contaminated with plutonium nitrate. The contaminated tissue was surgically excised and found to contain approximately 2.33 kBq of ^{239}Pu. A wound count performed four weeks later indicated that 59 Bq of ^{239}Pu still remained in the finger. Worksite personnel estimated systemic deposition of ^{239}Pu due to this accident as 85 Bq. This individual was medically treated with intravenous injections of Ca-DTPA. A total of 16 g Ca-DTPA was administered in 18 treatments during two months following the accident. Eighty-seven urine samples were collected and analysed over 14 years following the accident. This individual died 40 years post-intake at age 87.

2 Materials and Methods

Thirty-two soft tissue and 8 bone samples collected at autopsy were radiochemically analysed for ^{238}Pu, ^{239}Pu and ^{241}Am. A complete description of the radiochemical analysis protocol has been described elsewhere [1]. To estimate the plutonium intake, late urine measurements (past 100 days after the last Ca-DTPA injection), which were unaffected by chelation, and post-mortem radiochemical analysis results were evaluated using the IMBA Professional Plus® software [2].

* Corresponding author: m.avtandilashvili@wsu.edu

3 Results and Discussion

Post-mortem radiochemical analysis of autopsy tissues indicated that forty years post-accident 12.2±0.3 Bq of ^{239}Pu was retained in the liver and 17.5±0.7 Bq in the skeleton. Activity measured in the skin and muscle tissue sample from the wound site was 0.26±0.01 Bq while activity in the finger bone adjacent to the wound was measured as 1.09±0.03 Bq. Thus, a total of 1.35 Bq of ^{239}Pu was retained in the wound.

Activity in the lungs including thoracic lymph nodes was estimated to be 0.14±0.01 Bq, two orders of magnitude lower than activity in the liver. This observation confirmed the assumption of the soluble plutonium intake via wound injury to be the major source of internal contamination for this individual.

Application of the NCRP 156 wound model [3] with default parameters for soluble strong material resulted in a credible fit to the data ($p > 0.05$) (Fig. 1). The residual intake was estimated to be 47.6 Bq and estimated committed effective dose was 24.1 mSv. By accounting for ~77.8 Bq of ^{239}Pu excreted during Ca-DTPA treatment, the total intake was estimated to be 125 Bq. Without Ca-DTPA treatment, this individual would have received committed effective dose of 63.3 mSv. Chelation therapy reduced radiation dose by a factor of 2.6.

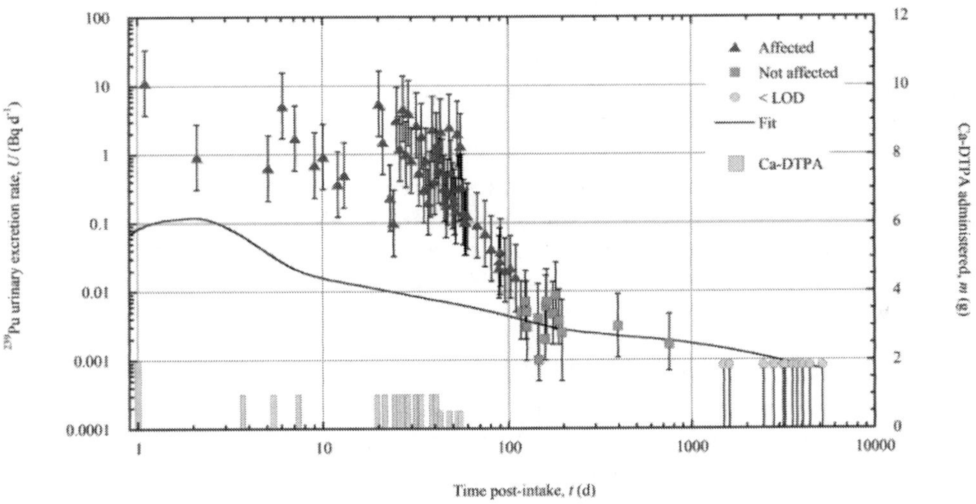

Fig. 1. ^{239}Pu daily urinary excretion and Ca-DTPA treatment data.

References

1. S. Y. Tolmachev, M. E. Ketterer, D. Hare, P. Doble, A. C. James. Proc Radiochimica Acta **1**, 173 (2011).
2. A. Birchall, M. Puncher, J. W. Marsh, K. Davis, M. R. Bailey, N. S. Jarvis, A. D. Peach, M. D. Dorrian, A. C. James. Radiat Prot Dosim **125**, 194 (2007).
3. National Council on Radiation Protection and Measurements. NCRP Report **156**, (2007).

Biokinetic analyses of the latest actinide model for internal dose assessment after an accidental inhalation at a nuclear facility in Oarai, Japan

Kotaro Tani[1,*], *Nobuhito* Ishigure[1], *Eunjoo* Kim[1], *Osamu* Kurihara[1]

[1] National Institutes for Quantum and Radiological Science and Technology (QST), National Institute of Radiological Sciences (NIRS), 4-9-1 Anagawa, Inage-ku 263-8555 Chiba, Japan

1 Introduction

An accidental inhalation of actinide nuclides, e.g. plutonium and americium, by 5 workers occurred due to an accident at Oarai Research and Development Institute, Japan Atomic Energy Agency in June 2017. Although wearing half-face masks, overalls and gloves, the workers possibly inhaled the radionuclides through tiny gaps between their faces and masks. *In vivo* counting by using a lung monitor and bioassay measurements for urinal and faecal samples have been performed at the National Institute of Radiological Sciences (NIRS) to assess their intake amounts and internal doses. Decorporation agents, i.e. Ca- and/or Zn-DTPA, were intravenously administered to the all workers for the purpose of enhancing the urinal excretion of actinide nuclides in blood, leading to difficulty in internal dose assessment because the biokinetics may deviate from the general model prediction. The effective dose of the worker who received the highest exposure is currently estimated, still as an initial assessment, to be less than 200 mSv based on previous biokinetic models recommended by International Commission on Radiological Protection (ICRP) [1-3]. The modified biokinetic models for plutonium [4, 5], however, had been proposed to reflect data become newly available since development of the previous models. Recently, ICRP also has published some relevant recommendations describing revised biokinetic models for several elements as the Occupational Intakes of Radionuclides (OIR) series [6-8]. In this study, as part of our effort on improvement of the internal dose assessment for the workers, biokinetic characteristics after inhalation of Pu-239 and Am-241 based on the latest models are compared with those based on the previous models.

2 Methods

Simultaneous differential equations written according to the previous/latest models shown in Table 1 were numerically analysed by using a commercial software, EQUATRAN-G (Omega Simulation, Japan). Absorption type of the respiratory models was selected to be Type M, and particle size in aerosol was selected to be 1, 5, 10, 15 or 20 μm in activity median aerodynamic diameter (AMAD). Relationships between AMAD and activity ratio, [(Excreted activity in faeces collected over first 5 days)/(Residual activity in the lung (the

* Corresponding author: tani.kotaro@qst.go.jp

thoracic airways and the thoracic lymph nodes) at third day after intake)], were compared between the previous and latest models.

Table 1. Previous and latest biokinetic models used for comparison.

Type of Biokinetic model	Previous model	Latest model
Respiratory tract model	ICRP Publ. 66 [2]	ICRP Publ. 130 (OIR part 1) [6]
Alimentary tract model	ICRP Publ. 30 [1]	ICRP Publ. 100 [9]
Systemic model	ICRP Publ. 67 [3]	Leggett et al. [5]

3 Results

Relationships between AMAD and the activity ratio are shown in Figure 1 (a) Pu-239 and (b) Am-241, resulting in large differences between the previous and latest models. This activity ratio is one of important facts, obtained from results of *in vivo* counting and bioassay measurements, to estimate AMAD, especially as in this case that information on particle size in aerosol is unavailable. The actual activity ratio of Am-241 measured after the accident was about 34, corresponding to more than 17 μm and 12 μm based on the previous and latest models. The results of this study demonstrate that AMAD is overestimated based on the previous models, leading to underestimation of actinide amount absorbed from respiratory tracts into blood because smaller particle in aerosol is likely to be inhaled deeply into the lungs. Therefore, it is clearly worth considering to apply the latest models in future studies for improvement of the internal dose assessment for the workers.

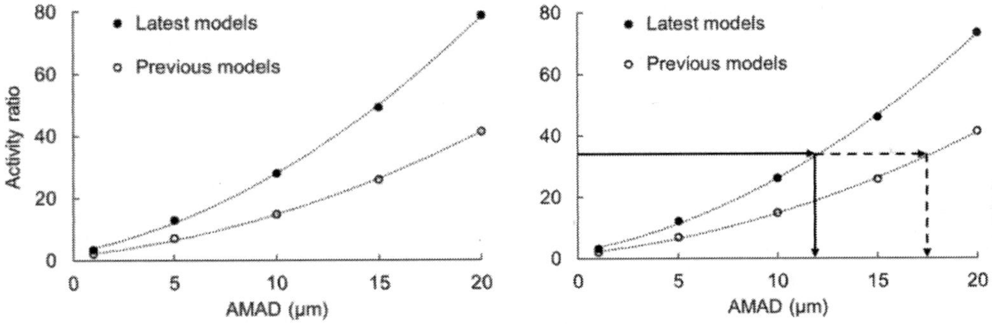

(a) Pu-239 (b) Am-241

Figure 1. Relationships between AMAD and activity ratio: [(Faecal activity for 5 days)/(Lung activity at third day)].

References

1. ICRP Publication 30 (Part 1), *Ann. ICRP* **2**, 3-4 (1979) ; 2. ICRP Publication 66, *Ann. ICRP* **24**, 1-3 (1994) ; 3. ICRP Publication 67, *Ann. ICRP* **23**, 3-4 (1993) ; 4. Leggett, R.W., *Radiat. Prot. Dosim.* **106**, 103-120 (2003) ; 5. Leggett, R.W. et al, *Radiat. Res.* **164**, 111-122 (2005) ; 6. ICRP Publication 130, *Ann. ICRP* **44**, 2 (2015) ; 7. ICRP Publication 134, *Ann. ICRP* **45**, 3-4 (2016) ; 8. ICRP Publication 137, *Ann. ICRP* **46**, 3-4 (2017) ; 9. ICRP Publication 100, *Ann. ICRP* **36**, 1-2 (2006).

BIO Web of Conferences **14**, 02010 (2019) https://doi.org/10.1051/bioconf/20191402010
HEIR 2018

Microdistribution of plutonium in human skeleton. Should we change the ICRP model?

Sergei Romanov[1,*], *Ekaterina* Lyovkina[1], *Sergei* Sypko[1], and *Aleksandr* Ephimov[1]

[1]Southern Urals Biophysics Institute, FMBA of Russia

Skeleton is one of the major organs of plutonium deposition within the human body. It is well known that plutonium distribution in bone tissue is quite uneven. According to generally accepted model, main fraction of plutonium activity resides on endosteal bone surface and Haversian canal walls. Present study is devoted to analyzing plutonium microdistribution in the thoracic vertebra obtained from a former Mayak PA worker and preserved by SUBI Human Tissue Repository; the analysis was carried out using neutron induced autoradiography.

Obtained results are in sharp contrast with the corresponding values for Leggett's biokinetic skeletal model recommended by ICRP, especially in case of cortical bone. According to main results of the study, the authors suggest adjusting the parameters of ICRP biokinetic model and to revise dose coefficients for radiation protection purposes.

* Corresponding author: romanov@subi.su

Inhomogeneous distribution of radon in different types of tissue in the human body

Andreas Maier[1,*], *Franziska* Papenfuß[1], *Michaela* Schmitt[1,2], *Claudia* Fournier[1] and *Gerhard* Kraft[1]

[1]GSI Helmholtz Centre, Biophysics Department, 64291 Darmstadt, Germany
[2]Goethe University Frankfurt, 60438 Frankfurt a. M., Germany

1 Introduction

For the therapy of inflammatory diseases of the musculoskeletal system, like rheumatoid arthritis or ankylosing spondylitis, radon (^{222}Rn) is a potential used modality [1]. On the other hand, the radioactive decay of radon accounts for ½ of the natural radiation exposure and is believed to be a major source for carcinogenesis [2].

The main dose is deposited in the lung, from which around 95% is originating from the short living radon decay products [3]. In order to understand the therapeutic effects of radon and the associated risk, knowledge of the distribution and the deposited dose in the human body is of crucial interest.

2 Material and methods

2.1 Solubility and distribution

In order to investigate the distribution of radon in the body, the solubility in different types of tissue is important. For the experiments, samples were exposed in a special designed radon exposure chamber with which conditions like in radon galleries can be simulated [4].

After a usual exposure time of one hour, the samples were removed and the γ-emitting decay products ^{214}Pb and ^{214}Bi were measured via γ-spectroscopy. From the variation in time of the measured activities, the initial radon activity concentration in the sample can be determined. By additional knowledge of the radon activity concentration during exposure, the radon solubility in the sample can be calculated.

For the experiments, different types of samples were chosen. Examples include pristine substances like oleic acid or linoleic acid, which are the most abundant fatty acids in the human body, or isotonic saline solution. Also tissue samples from dead pig like fat, muscle or bone were measured. Additionally the activities inside the body of a voluntary test person after radon therapy were determined at different body parts by γ-spectroscopy.

* Corresponding author: a.maier@gsi.de

2.2 Lung dosimetry

For the quantification of the dose, deposited in the lung, a mechanical lung model was developed. The radon decay products, which are the major contributor to the dose, are deposited on a glass fiber filter in a small tube. The flow through the model is regulated by a pump and measured by a flow-meter. This whole setup is placed in the radon exposure chamber and after the experiments, the filter is removed and the deposited activities were measured with the γ-detector. From these activities, the energy dose can be calculated [5].

3 Results and Outlook

The solubility of radon in fatty acids is about 50 times higher than in isotonic saline solution. A similar behaviour is found for fatty tissue in comparison to muscle tissue. For other tissues, experiments are currently under performance. The experiments with a voluntary test person just started but show promising results and will be continued.

The measured doses in the mechanical lung model are in the order of µGy or lower. At present, different parameters like the aerosol concentration are varied. A fraction of the radon decay products will attach to aerosols and thus change the deposition mechanism and location, depending on the size of the aerosol. In future experiments, a more anatomical correct lung model will be designed and a better variation of the aerosol concentration and size distribution will be established.

In our presentation, we want to present the methods and first results, giving promising indications for a better understanding of radon distribution and deposited dose in the human body.

We thank ACURADON Bad Kreuznach and Mrs. Brigitte Pfeiff for the possibility to measure in their radon gallery. This work is supported by the German Federal Ministry of Education and Research (BMBF) under project funding reference number 02NUK050A.

References

1. A. Van Tubergen, R. Landewé, D. Van Der Heijde, A. Hidding, N. Wolter, M. Asscher, A. Falkenbach, E. Genth, H.G. Thè, S. van der Linden, Arthritis Care Res. **45**, 430-438 (2001)
2. S. Darby, D. Hill, A. Auvinen, J. Barros-Dios, H. Baysson, F. Bochicchio, H. Deo, R. Falk, F. Forastiere, M. Hakama, BMJ **330**, 223-229 (2005)
3. ICRP, Occupational Intakes of Radionuclides: Part 3. ICRP Publication 137. Ann. ICRP 46(3/4) (2017)
4. A. Maier, P. van Beek, J. Hellmund, M. Durante, D. Schardt, G. Kraft, C. Fournier, Nucl. Instr. Meth. Phys. Res. B, **362**, 187-193 (2015)
5. A. Maier, S. Sternkopf, H. Schmidt-Böcking, G. Kraft, C. Fournier, GSI Scientific Report, 300 (2016)

Effective dose coefficients for inhaled radon and its progeny: ICRP's approach

James W. Marsh[1,*], *John D.* Harrison[1], *Dominique* Laurier[2], and *Margot* Tirmarche[3]

[1]Public Health England. Centre for Radiation, Chemical and Environmental Hazards, Chilton, Didcot, Oxon. OX11 0RQ, UK; PH (+44) 1235 831600; FAX (+44) 1235 833891
[2]Institut de Radioprotection et de Sûreté Nucléaire, Fontenay-aux-Roses, France
[3]Commissioner at ASN, Autorité de Sûreté Nucléaire, Montrouge, France

Abstract

The International Commission on Radiological Protection (ICRP) has recently published three reports on radon exposure: (i) Publication 115 on lung cancer risks from radon and radon progeny [1], (ii) Publication 126 on radiological protection against radon exposure [2] and (iii) Publication 137 on Occupational Intakes of Radionuclides (OIR), Part 3 [3]. The latter document gives doses coefficients for the inhalation of radon, thoron and their airborne progeny as well as recommendations for their use for the protection of workers. As with all other radionuclides, the effective dose coefficients are calculated with ICRP reference biokinetic and dosimetric models. Sufficient information and dosimetric data are given so that site-specific dose coefficients can be calculated based on measured aerosol parameter values.

In its review of epidemiological data on radon-induced lung cancer, ICRP concluded that the nominal risk coefficient for radon-induced lung cancer at low levels of exposure should be increased by about a factor of two in comparison to the value recommended previously (ICRP Publication 65) [1,4]. Accordingly, the upper reference level (URL) for radon in dwellings was reduced from 600 Bq m^{-3} to 300 Bq m^{-3}. As part of an integrated protection strategy, ICRP recommended the same URL of 300 Bq m^{-3} for workplaces [2].

Although protection against radon is based on measurement and control of levels of exposure, dose estimates are required in certain situations for workers. Based on both dosimetry and epidemiological data, the Commission recommends the use of a single dose coefficient of 3 mSv per mJ h m^{-3} (about 10 mSv per WLM) for the calculation of occupational doses following exposure to radon (^{222}Rn) progeny in underground mines and in building, in most circumstances. However, for indoor workplaces where workers are engaged in substantial physical activities and for workers in tourist caves, a dose coefficient of 6 mSv per mJ h m^{-3} (about 20 mSv per WLM) is consider to be more appropriate. In special cases, site-specific dose coefficients can be applied, if approved by the regulatory authority, where exposure conditions are non-typical, and sufficient, reliable aerosol data are available [3].

In this presentation, the application and use of dose coefficients for workplaces are discussed. Results of dose calculations for indoor workplaces, mines and tourist caves, are presented. The presentation also describes the general approach for the management of

* Corresponding author: James.Marsh@phe.gov.uk

radon exposure in workplaces based both on ICRP recommendations and the European directive (2013/59/EURATOM) [5].

References

1. International Commission on Radiological Protection. Lung cancer risk from radon and progeny and Statement on Radon. ICRP Publication 115, Ann. ICRP **40**(1) (2010).
2. International Commission on Radiological Protection. Radiological protection against radon exposure. ICRP Publication 126. Ann. ICRP **43**(3) (2014).
3. International Commission on Radiological Protection. Occupational intakes of radionuclides: Part 3. ICRP Publication 137. Ann. ICRP **46**(3/4). (2017).
4. International Commission on Radiological Protection. Protection against radon-222 at home and at work. ICRP Publication 65. Ann. ICRP **23** (2) (1993).
5. Council Directive 2013/59/Euratom of 5 December 2013. Official Journal of the European Union, vol. **57**, L13, (17 January 2014). ISSN 1977-0677.

In vivo screening measurements with common radiation protection instruments

Siria Medici[1,2,*], *Laurent* Desorgher[2], *Pierre* Carbonez[1,3], *Jérôme* Damet[1,2,3], *François* Bochud[2], and *Andreas* Pitzschke[2]

[1]European Organization for Nuclear Research (CERN), CH-1211 Geneva 23, Switzerland
[2]Institute of Radiation Physics, Lausanne University Hospital Center, Rue du Grand-Pré 1, CH-1007 Lausanne, Switzerland
[3]Department of Radiology, University of Otago, 2 Riccarton Ave, Christchurch, New Zealand

Professionals working with unsealed radioactive sources and/or in the presence of contaminated or activated materials are at risk of incorporation of radionuclides. Two scenarios of internal contamination are possible: 1) in the framework of isolated operations, such as the maintenance of activated accelerator components and radioactive waste handling, or 2) during routine operations, such as isotope collections and manipulation of radiopharmaceuticals.

Our work aims to set the basis for a new monitoring program for internal contamination at CERN (where both exposure scenarios are present) by using conventional radiation protection instruments.

The monitoring program is inspired by the Swiss regulation and consists in screening measurements that aim to determine if a significant internal contamination occurred. The screening measurements are quick and simple and can usually be performed by the workers themselves. If a radionuclide-specific threshold is exceeded during a screening measurement, the worker must undergo a specific incorporation measurement to assess the intake activity that is then used for the evaluation of the committed dose. If the screening measurement is negative, no dose assessment is required. Therefore, the screening thresholds are set in order to guarantee that the committed effective dose does not exceed 1 mSv/year.

In our case, the screening measurements are performed as soon as possible after an isolated operation or periodically in the case of routine operations.

The development of the monitoring program includes the characterisation of conventional radiation protection instruments such as dose rate meters, contamination monitors and probes, to determine whether they are suitable to perform screening measurements. The characterisation is done using different isotopes to cover a wide photon energy range (124 keV - 1.3 MeV). Instruments' characteristic limits are calculated accordingly to the ISO 11929 standard [1] and are provided for different background radiation levels. Simple and easily reproducible phantoms are used to link the measured quantity to the incorporated activity. The obtained experimental results are cross-validated with GEANT4 Monte Carlo simulations [2]. The minimum detectable activity is derived from the instrument's

* Corresponding author: siria.maria.chiara.medici@cern.ch

characteristic limits. The biokinetic models recently published by the International Commission on Radiological Protection [3] are used to assess:
1) the minimum detectable committed effective dose according to the time elapsed between the manipulation and the measurement in the case of isolated operations,
2) the optimal compromise between screening threshold and screening interval duration in the case of routine operations.

Our results show that, for a given radionuclide inserted into the phantoms, the response of the different instruments varies up to one order of magnitude. Therefore, the instrument choice is crucial and has to be done according to the radionuclide that is going to be detected.

The suggested screening measurements allow to detect the presence of an internal contamination using conventional radiation protection instruments. For the radionuclides used during the characterisation, a screening measurement performed the day following an isolated operation allows the detection of an internal contamination leading to a committed effective dose in the order of tens up to a few hundreds of μSv, depending on the chosen instrument.

The methodology provides local radiation protection experts with the necessary tools to set a dedicated monitoring program for internal contamination.

In this presentation, we will explain the methodology used to implement in vivo screening measurements at CERN and show that it can be easily adapted to the needs of other research centres. We will present the performances of six commonly used radiation protection instruments and illustrate how to use them to perform screening measurements considering examples of real case exposure scenarios.

References

1. International Organization for Standardization, ISO Standard 11929 (2010).
2. J. Allison, et al, Nucl Instrum Methods Phys Res A, **835**, 186-225 (2016).
3. International Commission on Radiological Protection, ICRP Publication 130, Ann. ICRP **44**(2) (2015).

Evaluation of internal exposure in French nuclear power plants based on whole body counting

Mathieu Sailly[1*], *Bernard* Landry[2], and *Isabelle* Le-Couteulx[3]

[1] CHU Rouen, 76000 Rouen, France
[2] EDF CNPE du Blayais, BP 27, 33820 Saint Ciers sur Gironde, France
[3] EDF CNPE de Paluel, BP 48, 76450 Cany Barville, France

Introduction

At EDF, dose assessments for internal contamination were so far mainly done by biologists using in vitro measurements. Several factors lead to change this practice: French regulations which attribute this assessment to occupational physicians [1], accreditation of whole body measurements which legitimates the use of this type of analysis and the development of Cador, a home-made Excel file [2] which allows to easily perform the assessment. The objective of this work was to estimate the interest of whole body measurements for internal contaminations.

Material and method

An analysis of 149 cases of internal contamination which occurred in French nuclear power plants between 2008 and 2015 was conducted. More than one thousand whole body counting spectrums were treated with double-checking. The analysis was led by the last version of Cador using the French Recommendation of HAS [3] and the Chi-squared test provided by European IDEAS guidelines [4].

Most of the contaminations were situated between 0.1 mSv and 1 mSv. This level corresponds to the intermediate level of the French Recommendations of Good Practice and to the level 1 of the IDEAS guidelines. For this level, an estimation based on a single type of analysis is considered sufficient. In this respect, the assessment might be done with whole body measurements only.

In the Recommendations of Good Practice (RGP) the goodness of fit is considered acceptable if there is no more than a factor 3 between the various estimates of intake. The uncertainties are not taken into account. In the IDEAS guidelines, the goodness of fit between the measures and the model is assessed against a Chi-squared test. This method requires knowing the uncertainties of measures. IDEAS guidelines provide values of

* Corresponding author : matsailly@icloud.com

geometric standard deviation, so called Scattering factors, for various analyses. For whole body measurements of high energy radiation, the Scattering Factor is 1.2.

The first stage refers to RGP method and consists in looking at the gap between the various estimates. The second stage refers to IDEAS method with the Chi-squared test. A coloured rectangle summarizes the results of both methods. The rectangle is green, orange or red depending on whether the fit is good with respectively both methods, any of them or none of them.

Results

30 cases could not be analysed because of an insufficient number of whole body measures. Among 119 remaining cases, 21 cases corresponded to an inhalation model of hot particles where the assessment of activity needed another method [5]. For 32 cases, the activity was too low or the number of measures was insufficient.

For the 66 remaining cases, we were able to exactly identify an incorporation model described by ICRP. RGP and Chi-squared tests were valid in 79% of cases and RGP alone in remaining 21%. The identified models and AMAD observed are shown in table 1.
In an isolated way or not, we found Cobalt 58 in 55% of the cases, Cobalt 60 in 48% of the cases and Argent 110m in 26% of the cases. We also observed Caesium 137, Antimony 124, Zirconium-Niobium 95, Chromium 51 and Manganese 54.

Inhalation : 60 cases (91%)			**Ingestion** : 6 cases (9%)
5 µm AMAD	1 µm AMAD	1 or 5 µm AMAD + hot particles	
30 cases (46%)	22 cases (33%)	8 cases (12%)	

Table 1: models identified by RGP and chi-squared test

Except the cases of inhalation of a hot particle, the evaluation of the incorporated activity by whole body measurements was comparable to faeces analysis in half of cases. In the other half, the estimated activity from faecal examinations was lower than activity estimated by the whole-body measures.

Conclusion

Finally, 34 cases resulted in a committed effective dose superior to 0,1mSv but lower than 0.5mSv. The whole-body measurements allowed the dose assessment if the time of counting and the number of measure are sufficient.

Key words: Occupational medicine, radiation protection, nuclear power plant, internal contamination, whole body measurements, committed effective dose.

References

1. JORF n°0181 du 6 août 2013 page 13401 texte n° 34. Arrêté du 17 juillet 2013 relatif à la carte de suivi médical et au suivi dosimétrique des travailleurs exposés aux rayonnements ionisants, https://www.legifrance.gouv.fr/jo_pdf.do?id=JORFTEXT000027806111
2. Landry B, Un outil de calcul de la dose efficace engagée, 2018 Radioprotection 53(1) : 33-44
3. Recommandations de Bonne Pratique. Surveillance médico-professionnelle de l'exposition interne aux radionucléides en Installations Nucléaires de Base, juillet 2011, www.chu-rouen.fr/sfmt/pages/Recommandations.php
4. Castellani CM, Marsh JW, Hurtgen C, Blanchardon E, Berard P, Giussani A, Lopez MA. IDEAS Guidelines (Version 2) for the Estimation of Comitted Doses from Incorporation Monitoring Data. EURADOS, 2013. Braunschweig, March 2013, http://www.eurados.org/~/media/Files/Eurados/documents/EURADOS Report 2013-01 online version.pdf
5. Sailly M, Lecouteulx I, Rollin L. Evaluation de l'activité d'une particule chaude radioactive incorporée chez un travailleur du nucléaire. 2018
6. Sailly M, Etude de la cinétique anthropogammamétrique après contamination interne par radionucléides émetteurs gamma en installation nucléaire de base, doctoral thesis to be published.

Comprehensive dosimetry for seven exposure sources at the earliest US uranium processing facility

Elizabeth D. Ellis[1,*], *Richard W.* Leggett[2], *Ashley P.* Golden[1], and *John D.* Boice, Jr[3]

[1]Oak Ridge Institute for Science and Education, 100 ORAU Way, Oak Ridge, TN 37831, USA
[2]Oak Ridge National Laboratory, Oak Ridge, TN 37830, USA
[3]National Council on Radiation Protection and Measurements, 7910 Woodmont Ave, Bethesda, MD 20814, USA; Vanderbilt University Medical School, 2525 West End Ave, Nashville, TN 37203, USA

Abstract

Mallinckrodt Chemical Works (MCW) was the earliest uranium processing facility in the United States, beginning in 1942. The 2,514 workers included in the epidemiologic study were exposed to external gamma radiation, medical x-rays, internal radiation from intakes of pitchblende ore and its extracted radionuclides (mainly uranium isotopes and radium-226), and ambient levels of radon and its progeny [1].

Multiple sources of exposure data were available to a degree unique for an epidemiologic study [2]. Personal film badges captured external exposures in 12,686 records. Additional external exposure included over 31,297 occupational medical x-rays and 210 exposure records from employment outside of MCW. Organ dose estimates were adjusted for photon energy and job-specific orientation of workers to the radiation source during processing. Intakes of uranium and radium were based on 39,451 uranium urine bioassays and 2,341 breath radon measurements using International Commission on Radiological Protection (ICRP) biokinetic models to estimate organ-specific radiation absorbed dose [3]. Estimates of exposure to airborne radon and its progeny were available from 6,846 ambient radon measurements in work areas where radium-containing material was handled or stored, together with estimated exposure times in these areas based on job titles. Dose estimates for radon progeny were based on models and methods recently recommended by ICRP [4]. Contributions of uranium dust and silica exposures to risk were estimated from 6,300gol worker-specific estimates of daily exposure to uranium dust [2]. These data were used to calculate job-specific daily time-weighted average dust exposures. Because intake of uranium is best described by bioassay data, the dust exposures were used to explore potential associations with heavy metal toxicity and silica, independent of the radioactive content.

This comprehensive dosimetric approach follows methods outlined by the National Council on Radiation Protection Scientific Report 178 for the Million Worker Study [5]. Annual doses were calculated for 6 organs: lung, brain, heart, kidney, colon and red bone marrow. Mean and median dose estimates for external, internal, and total radiation exposure for each organ of interest are listed in Table 1.

* Corresponding author: betsy.ellis@orau.org

Table 1. Summary of contributions of each exposure source to organ-specific doses

Organ	External-Film Badge/TLD N=2,514		External-Medical x-ray N=2,514		Internal U and Ra N=1,886		Total External + Internal N=2,514	
	Mean dose in mGy [Std] Median dose in mGy [Max]							
Lung	36.0	[68.9]	10.7	[8.7]	27.1	[39.2]	69.9	[103.0]
	11.2	[724.2]	7.2	[53.6]	11.2	[416.3]	33.1	[885.2]
Brain	33.0	[62.5]	4.2	[3.6]	0.079	[0.1]	37.2	[63.9]
	10.7	[738.4]	3.0	[20.0]	0.040	[1.4]	15.2	[749.6]
Heart	35.6	[68.4]	11.8	[9.7]	0.089	[0.1]	47.5	[73.0]
	11.0	[710.8]	8.0	[59.3]	0.044	[1.4]	23.3	[737.6]
Kidney	27.9	[53.1]	10.7	[8.7]	0.73	[0.9]	39.2	[57.9]
	8.9	[603.5]	7.2	[53.6]	0.41	[11.6]	20.2	[637.5]
Colon	35.6	[68.4]	2.3	[1.9]	0.13	[0.2]	38.0	[69.4]
	11.0	[711.5]	1.8	[11.9]	0.05	[2.6]	13.4	[713.5]
Bone	32.9	[62.9]	2.2	[1.9]	0.57	[1.1]	35.6	[64.3]
marrow	10.3	[675.5]	1.8	[11.1]	0.18	[10.3]	12.9	[682.7]

The largest organ doses were to the lung (mean 69.9 mGy; max 885 mGy) and heart (mean 47.5 mGy; max 737.6 mGy). External gamma dose was the largest contributor to dose for all organs, but doses from medical x-rays were also important. Average annual dust exposure was 15.1 mg/m^3 (max 400 mg/m^3).

The extensive collection of radiation exposure data available for the MCW workers provides the broadest application to date of the methodology endorsed by the NCRP [5] for construction of organ doses for use in epidemiologic studies. Such comprehensive dosimetry is required to produce the best radiation risk estimates for addressing current radiation issues on the effectiveness of low dose-rate exposures. These estimates are relevant to: projecting risks from environmental contamination from nuclear waste, terrorist events, and radiations used in medical diagnosis or treatment; compensation issues related to prior exposures; improved protection standards for workers and the public; evaluation of the safety and radiation risks for astronauts leaving Earth's orbit; and appropriateness of the linear non-threshold model as a principle for radiation protection guidance.

References

1. J.D Boice Jr., E.D. Ellis, A.P. Golden, D.J. Girardi, S.S. Cohen, H. Chen, M.T. Mumma, R.E. Shore, R.W. Leggett. Health Phys., **114**(4): 381-385 (2018).
2. E.D. Ellis, J.D. Boice Jr., A.P. Golden, D.J. Girardi, S.S. Cohen, M.T. Mumma, R.E. Shore, R.W. Leggett, G.D. Kerr. Health Phys., **114**(4): 386-297 (2018).
3. International Commission on Radiological Protection. Dose coefficients for intakes of radionuclides by workers: Part 4. New York: Elsevier; ICRP Publication 68, *Ann ICRP* 24(4) (1994).
4. International Commission on Radiological Protection. Occupational intakes of radionuclides: Part 3. London: Sage; ICRP Publication 137, *Ann ICRP* 46(3-4) (2017).
5. National Council on Radiation Protection and Measurements. Deriving organ doses and their uncertainty for epidemiologic studies (with a focus on the One Million U.S. Workers and Veterans Study of Low-dose Radiation Health Effects). Bethesda, MD; NCRP Report No. 178 (2018).

BIO Web of Conferences **14**, 03006 (2019) https://doi.org/10.1051/bioconf/20191403006
HEIR 2018

Dose Assessment of workers long term exposed to chronic intakes of enriched uranium

María Antonia López[1,*], *Soledad* Sierra[2], *Inmaculada* Sierra[1], *Carolina* Hernández[1], *Daniel* Garcia[2], *Agustín* Pérez[2]

[1]CIEMAT, Internal Dosimetry, Avda.Complutense 40, 28040 Madrid, Spain
[2]ENUSA, Fábrica de elementos combustibles, Carreterade Salamanca a Ledesma km26, 37115 Juzbado (Salamanca), Spain

CIEMAT Internal Dosimetry Service started in 2014 an individual monitoring program of in vitro bioassay measurements of low enriched uranium in urine by radiochemical procedure and alpha spectrometry, for workers involved in the fabrication of nuclear fuel elements for Spanish Nuclear Power Plants.

A total of 198 workers at risk of internal exposures in the Juzbado facility of ENUSA in Salamanca (Spain) have been included in routine, task-related or confirmatory monitoring programs, providing 24h urine samples for the determination of the daily excreted activity of ^{234}U, ^{238}U and ^{235}U, with a Minimum Detectable Amount (MDA) of 0.5 mBq/d for each isotope. A "blank" urine sample of the new employees at the ENUSA workplace is obtained prior to start working in areas at risk of internal exposure for background subtraction. A study of the dietary contribution of natural uranium in the area is in process.

The monitoring data of 72 workers obtained in 2014 suggested a scenario of chronic intakes through inhalation of low enriched uranium at Juzbado facility. Currently workers identified at risk of internal exposure in the plant are monitored in routine at CIEMAT Bioassay laboratory in Madrid, 250 km from Juzbado. Excreta samples preserved in nitric acid are transported from Salamanca to Madrid. A creatinine test for the validation of 24h excretion is performed and correction of daily activity rate is carried out accordingly. Taking into account the available equipment and human resources of CIEMAT internal dosimetry service for measurements and dose assessments, frequency of monitoring was established according to first results of Committed Effective Dose E(50) obtained in 2014, being annual monitoring in case of E(50)< 5 mSv and semi-annual monitoring when E(50)>5 mSv.

CIEMAT established a methodology for dose evaluation based on IDEAS Guidelines [1] and ISO 16638-Part 1 [2], using IMBA software (v 4.1.60). Chronic inhalation of low enriched uranium since the beginning of internal exposure of each worker at the Juzbado fabric was assumed. The intake scenario consists in Uranium oxides, type S of solubility and a particle size of 5 μm of AMAD (default value recommended by ICRP). Dose assessment was carried out in an ICRP78/ICRP119 frame for occupational exposures.
Interpretation of monitoring data takes account of type A and type B uncertainties by calculating the corresponding Scattering Factor [1] of each measurement result, for intake estimation. The Chi squares test is used for confirming the goodness of fit of the monitoring data with the prediction of the excretion model for the assessed intake and permits to corroborate the conditions and assumptions of the internal exposure at the workplace.

*Corresponding author: ma.lopez@ciemat.es

Intake and dose E(50) are obtained separately for ^{234}U and ^{238}U when activity concentration (Bq/d) is above MDA. Total Intake of enriched uranium is estimated using isotopic composition of low enriched uranium by ISO 16638 Part 1[2] (82% alpha activity corresponds to ^{234}U) for calculating the intakes of uranium isotopes whose data are below MDA. Dose contributions of annual chronic intakes were assessed for 106 workers with results > MDA. Results of E(50) > 1 mSv/year are reported to workers. In case of partial annual exposures, doses of E(50) > 0.2 mSv are also reported.

This publication focuses on the group of 37 workers chronically exposed to the inhalation of enriched uranium during 30 years. Bioassay monitoring data obtained in the period 2014-2018 resulted in daily excreted activities in the range of 1-15 mBq/d of ^{234}U in urine for this long term exposed staff. Isotopic composition of excreta samples confirmed the intakes of low enriched uranium, with an average $^{234}U/^{238}U$ ratio of 4.4 of urine activity rate. Annual chronic intakes with doses E(50) above 1 mSv/year were detected in 50% of the cases (see Figure 1). Maximum annual intake corresponds to one worker exposed since 1985, with E(50) = 5 mSv/year.

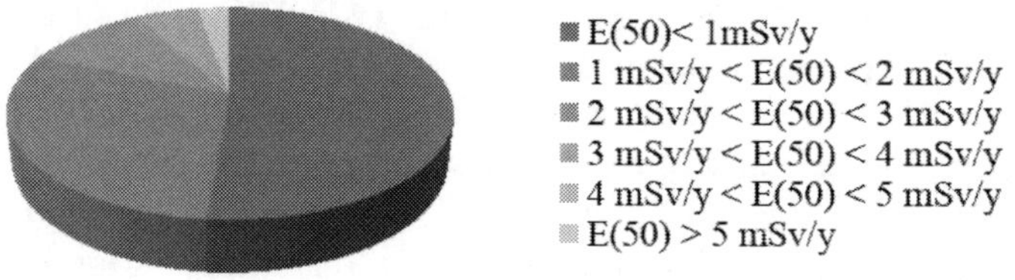

Figure 1.- Dose distribution of long term (30 years) exposed workers to chronic inhalation of low enriched uranium

Highest intakes correspond to workers developing same specific tasks in the process of the fabrication of nuclear fuel elements for nuclear power plants with enriched uranium. Dosimetry information is contributing to improve the radiation protection conditions of the nuclear workforce in Juzbado.

References

1.- C.M. Castellani, J.W. Marsh, C. Hurtgen, E. Blanchardon, P. Berard, A. Giussani, M.A. Lopez, *IDEAS Guidelines (Version 2) for the Estimation of Committed Doses from Incorporation Monitoring Data.* EURADOS Report 2013-01.Braunschweig (2013).

2.-International Organization for Standardization. ISO 16638-Part 1 Radiation Protection: Monitoring and internal dosimetry for specific materials. Part 1: Uranium. ISO: Geneva (2015).

Internal dose assessment for the French cohort of uranium workers

Estelle Davesne[1,*], *Olivier* Laurent[2], *Eric* Samson[2], *Iris* Jovanovic[3], *Ségolène* Bouet[2], *Cécile* Challeton-de Vathaire[1], and *Eric* Blanchardon[1]

[1]Institut de Radioprotection et de Sûreté Nucléaire (IRSN), PSE-Sante/SDOS/LEDI, 92262 Fontenay-aux-Roses Cedex, France
[2] Institut de Radioprotection et de Sûreté Nucléaire (IRSN), PSE-Sante/SESANE/LEPID, 92262 Fontenay-aux-Roses Cedex, France
[3] Amplexor, Montigny-le-Bretonneux, France

Recently, several epidemiological studies[1-14] were dedicated to identifying potential health effects from incorporated radionuclides. To do so, the exposure and/or dose of individuals were quantified and associations with health status investigated. Since nuclear workers' health and occupational exposure are carefully monitored, they form a particularly interesting population to study risks potentially induced by internal contamination. To reveal any dose-response relation, lifetime internal dose must be assessed for all workers of the epidemiological cohort.

For this, the measurement results of retained and excreted activity (bioassay) are interpreted with biokinetic and dosimetric models. The models best adapted to the situation are chosen on the basis of known or assumed conditions of exposure: radionuclide(s), isotopic ratios, intake time(s), intake route(s): inhalation, ingestion, wound; physicochemical properties of the radioactive material: solubility and activity median aerodynamic diameter (AMAD) for an aerosol. Some or all of these parameters are, in general, unknown and often difficult to characterize regarding past intakes in which epidemiological studies are interested.

In a subset of the French cohort of workers exposed to uranium during nuclear fuel cycle operations (TRACY, [3; 15]), the exposure conditions are documented by job exposure matrices (JEM) relating workplaces and periods of exposure with chemical forms and isotopic compositions of handled uranium compounds, and by incident registries.

From bioassay data, incident registry, JEM and administrative files, annual absorbed doses received by different organs were estimated for these workers through a dosimetric protocol implemented in the DOSEPI software. DOSEPI allowed the estimation of lifetime doses for more than 2,800 workers in 2.5 hours of computation on a supercomputer, from an Excel® file of bioassay results and exposure conditions translated into input files by Visual Basic® macro procedures.

DOSEPI will be presented in details along with the dosimetric protocol on which it is based. Doses estimated for the TRACY cohort subset will also be presented and the application to other cohorts and exposures will be discussed.

* Corresponding author: estelle.davesne@irsn.fr

References

1. T.V. Azizova, E. Batistatou, E.S. Grigorieva, et al., Radiation Research **189**, 371-388 (2018)
2. J.D. Boice, Jr., S.S. Cohen, M.T. Mumma, et al., J Radiol Prot **28**, 303-325 (2008)
3. S. Bouet, E. Davesne, E. Samson, et al., International archives of occupational and environmental health (submitted)
4. E.S. Gilbert, M.E. Sokolnikov, D.L. Preston, et al., Radiation research **179**, 332-342 (2013)
5. M. Gillies, I. Kuznetsova, M. Sokolnikov, et al., Radiation Research **188**, 645-660 (2017)
6. M. Kreuzer, F. Dufey, D. Laurier, et al., International archives of occupational and environmental health **88**, 431-441 (2015)
7. M. Kreuzer, C. Sobotzki, N. Fenske, et al., Occupational and environmental medicine **74**, 252-258 (2017)
8. I.S. Kuznetsova, E.V. Labutina, N. Hunter, PloS one **11**, e0162710 (2016)
9. E. Rage, S. Caer-Lorho, D. Drubay, et al., International archives of occupational and environmental health **88**, 717-730 (2015)
10. E. Samson, K. Leuraud, E. Rage, et al., Radioprotection (Submitted)
11. J.H. Yiin, J.L. Anderson, R.D. Daniels, et al., American journal of industrial medicine **60**, 96-108 (2017)
12. L.B. Zablotska, N. Fenske, M. Schnelzer, et al., International archives of occupational and environmental health **91**, 91-103 (2018)
13. S. Zhivin, I. Guseva Canu, E. Davesne, et al., Occupational and environmental medicine **75**, 270-276 (2018)
14. J. Grellier, W. Atkinson, P. Berard, et al., Epidemiology **28**, 675-684 (2017)
15. E. Samson, I. Piot, S. Zhivin, et al., BMJ open **6**, e010316 (2016)

Case studies in brain dosimetry for internal emitters: Is more detail needed for epidemiology?

Richard W. Leggett[1,*], Sergei Y. Tolmachev[2], John D. Boice, Jr[3]

[1]Oak Ridge National Laboratory, Oak Ridge, TN 37831, USA
[2]U.S. Transuranium and Uranium Registries, College of Pharmacy, Washington State University, 1845 Terminal Drive, Suite 201, Richland, WA 99354, USA
[3]National Council on Radiation Protection and Measurements, 7910 Woodmont Ave, Ste 400, Bethesda, MD 20814, USA

Abstract

Element-specific biokinetic models are used to reconstruct doses to systemic tissues from internal emitters. These models typically depict explicitly only those tissues that tend to dominate the systemic behaviour of the element over time. The remaining tissues are aggregated into a pool called *Other tissue* in which activity is assumed to be uniformly distributed. Explicitly identified tissues usually consist of some subset of the tissues liver, kidneys, bone, bone marrow, gonads, thyroid, spleen, and skin.

The brain is included explicitly in systemic biokinetic models for a few elements but typically is addressed as an implicit mass fraction of *Other tissue*. There is increasing interest in the potential adverse effects of internal emitters, particularly alpha emitters, on the brain as limited analogues for galactic cosmic ray (GCR) exposures during space travel and for possible assessment of radiogenic effects on brain in nuclear medicine patients and radiation workers. For National Aeronautics and Space Administration, the need is to provide protection against in-flight behavioural and cognitive impairments from GCRs on the central nervous system, as well as long-term dementia and motor neuron diseases [1,2]. The Million Worker Study (MWS), underway in the US, is estimating brain doses from exposure to radionuclides and evaluating dementia, Alzheimer's disease, Parkinson's disease, and motor neuron disease as possible adverse outcomes of combined high- and low-LET exposures to brain tissue [2,3].

This paper summarizes an assessment of potential improvements in brain dosimetry for internal emitters from explicit modelling of brain kinetics in place of treating the brain as an implicit mass fraction of *Other tissue*. Comparisons are made of dose coefficients for selected radionuclides based on alternate versions of the systemic biokinetic model for each radionuclide, differing only in the handling of brain tissue.

As an illustration, the systemic model for plutonium (Pu) used in the MWS [4] includes brain implicitly in *Other tissue*. The most relevant brain-specific data available for modelling brain kinetics of Pu appears to be autopsy data for Pu workers. Since 1968, the U.S. Transuranium and Uranium Registries (USTUR) has studied the biokinetics (deposition, translocation, retention, and excretion) and tissue dosimetry of actinide elements in occupationally exposed individuals (volunteer Registrants) [5]. The USTUR holds data on work history, radiation exposure and bioassay measurements, and medical records from more than 400 former nuclear workers, mainly exposed to ^{239}Pu. The activity of ^{239}Pu in brain has been measured post mortem in several individuals. As a central estimate (either mean or median) for these individuals, the brain contains ~0.2% as much ^{239}Pu as liver and skeleton combined (Fig. 1). Based on USTUR data, a single compartment

* Corresponding author: leggettrw@ornl.gov

representing brain was added to the Pu model, and parameter values for brain were set to yield a long-term total activity ratio *Brain / (Liver + Skeleton)* of 0.002.

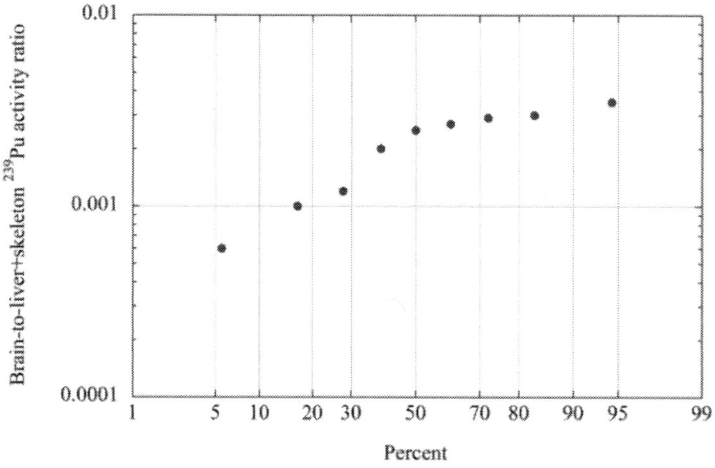

Fig. 1. Brain activity of ^{239}Pu relative to activity in liver plus skeleton.

Table 1 compares dose coefficients for brain for ^{239}Pu and other radionuclides based on alternate versions of the systemic biokinetic model for each radionuclide: Version A which includes brain implicitly in *Other tissue* and Version B which explicitly depicts a brain pool with kinetics based on brain-specific radiobiological data. Acute input of the radionuclide into blood is assumed. As illustrated in Table 1, results of the study to this point suggest that explicit biokinetic modelling of a brain pool for elements of interest is likely to result in a moderate increase in estimated dose to the brain from most internal emitters.

Table 1. Comparison of dose coefficients (Sv Bq^{-1}) for brain for acute input to blood, based on biokinetic model with brain included in *Other tissue* and modified version with an explicit brain pool.

Nuclide	Version of biokinetic model		Ratio B:A
	(A) Implicit brain	(B) Explicit brain	
^{134}Cs	5.9×10^{-9}	7.6×10^{-9}	1.3
^{54}Mn	1.4×10^{-9}	2.4×10^{-9}	1.7
^{203}Hg vapor	6.4×10^{-10}	9.0×10^{-10}	1.4
^{210}Po	3.1×10^{-7}	5.4×10^{-7}	1.7
^{239}Pu	2.2×10^{-5}	2.6×10^{-5}	1.2

References

1. National Council on Radiation Protection and Measurements. NCRP Commentary No. **25** (2016)
2. J. D. Boice JD Jr. Health Phys **112**,392 (2017).
3. J. D. Boice, E. D. Ellis, A. P. Golden, D. J. Girardi, S. S. Cohen, H. Chen, M. T. Mumma, R. E. Shore, R. W. Leggett. Health Phys. **114**, 381 (2018).
4. R. W. Leggett, K. F. Eckerman, V. F. Khokhryakov, K. G. Suslova, M. P. Krahenbuhl S. C. Miller. Radiat. Res. **164**, 111 (2005).
5. The United States Transuranium and Uranium Registries. www.ustur.wsu.edu

Uncertainties in radiation exposure assessment in the Wismut cohort: a preliminary evaluation

Veronika Deffner[1,2,*], *Michaela* Kreuzer[1], *Christina* Sobotzki[1], *Matthias* Aßenmacher[2], *Denise* Güthlin[3], *Christian* Kaiser[3], *Helmut* Küchenhoff[2], and *Nora* Fenske[1]

[1]Federal Office for Radiation Protection (BfS), Division Effects and Risks of Ionising and Non-Ionising Radiation, Neuherberg, Germany
[2]Statistical Consulting Unit StaBLab, Department of Statistics, LMU Munich, Germany
[3] Helmholtz Zentrum München, Department of Radiation Sciences, Institute of Radiation Protection, Neuherberg, Germany

1 Background

The German "Wismut" cohort is one of the largest cohort studies of uranium miners who have been occupationally exposed to radon [1]; the cohort includes 58,974 males accumulating over 2 million person years at risk (follow-up period 1946-2013). A job-exposure matrix (JEM, [2, 3]) was created for exposures to radon and its progeny, external gamma radiation and long-lived radionuclides. According to the availability of radiation measurements, different approaches of exposure assessment were developed. Individual radiation exposure in the Wismut cohort was calculated based on the JEM and individual job histories.

The elaborated exposure assessment procedure provides well-grounded exposure estimates, but may involve potential uncertainties. Numerous risk analyses for lung cancer mortality (e.g. [4]) and other diseases have been performed depending on radiation exposure, but so far without adjustment for such uncertainties. This work is a first step towards accounting for uncertainties in radon exposure in risk analyses for lung cancer mortality in the Wismut cohort. Potential sources of uncertainty were identified and preliminarily evaluated.

2 Identification and preliminary evaluation of the sources of uncertainties in exposure assessment

The exposure assessment procedure for the Wismut cohort can be structured as shown in Figure 1, pointing out two main types of uncertainty.

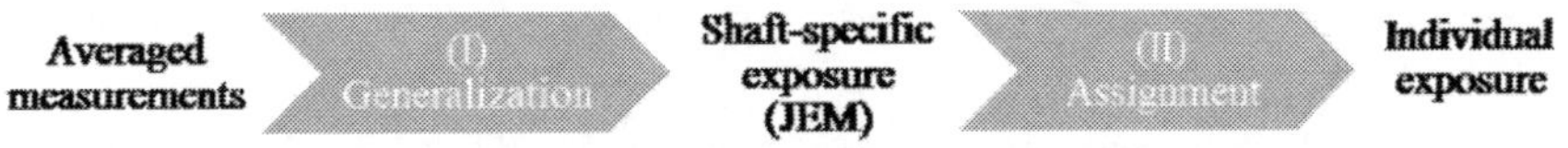

Fig. 1. Main steps in exposure assessment for the Wismut cohort.

The structure of uncertainties is complex because the multi-stage exposure assessment varies over time and depends on the working conditions and thus, involves different types

* Corresponding author: vdeffner@bfs.de

and sizes of errors. Errors in the exposure assessment may arise from the generalization of single exposure measurements (generalization error), from the assignment of general exposure estimates to single miners (assignment error) and from estimation errors (procedural measurement error, documentation error, parameter uncertainties, experts' evaluation error, transfer error, approximation error) in all stages of the exposure estimation process. The preliminary evaluation of the relevance of the uncertainties is shown in Table 1.

Tab. 1. Preliminary evaluation of the potential sources of uncertainties in the exposure assessment for the Wismut cohort.

Type of uncertainty	Details	Relevance
Generalization error	Usage of averages of single values for shaft-specific exposure assessment	major
Parameter uncertainties	Uncertainties in the determination of parameters, e.g. evaluation factor, activity factor	major
Assignment error	Assignment of group-specific values to an individual	medium
Transfer error	Data transfer to another calendar year or object	medium
Documentation error	Documentation of measurements and occupational histories	medium-minor
Experts' evaluation error	Determination of exposure values by experts	minor
Procedural measurement error	Human and technical errors in the measurement procedure	minor
Approximation error	Approximation through estimation equations and rounding	minor

These findings contribute towards the specification of a measurement error model, which is necessary for the estimation of lung cancer risk depending on radon exposure taking measurement error into account.

References

1. M. Kreuzer, M. Schnelzer, A. Tschense, L. Walsh, B. Grosche B, Int. J. Epidemiol. **39**, 980–987 (2010)
2. F. Lehmann, L. Hambeck, K.H. Linkert, H. Lutze, H. Meyer, H. Reiber, A. Reinisch. H.-J. Renner, T. Seifert, F. Wolf, *Belastung durch ionisierende Strahlung im Uranerzbergbau der ehemaligen DDR: Abschlußbericht zu einem Forschungsvorhaben* (1998)
3. F. Lehmann, *Job-Exposure-Matrix "Ionisierende Strahlung im Uranerzbergbau der ehemaligen DDR." Technical Report, Version 06/2004* (2004)
4. M. Kreuzer, C. Sobotzki, M. Schnelzer, N. Fenske, Radiat. Res., **189**, 165-176 (2018)

https://doi.org/10.1051/bioconf/20191403010

A reliable and robust method for monitoring large populations to assess thyroid internal exposure in a nuclear accident: A proposal based on experiences from Fukushima

Kazuaki Yajima[1], *Kim* Eunjoo[1], *Kotaro* Tani[1], *Hideo* Tatsuzaki[1], and *Osamu* Kurihara[1,*]

[1]National Institutes for Quantum and Radiological Science and Technology, National Institute of Radiological Sciences (QST-NIRS), 4-9-1 Anagawa, Chiba-city, Chiba, Japan

1 Introduction

The system for nuclear emergency response in Japan has dramatically changed since the Fukushima Daiichi Nuclear Power Plant (FDNPP) accident in 2011. In September 2012, the Nuclear Regulation Authority (NRA) of Japan was established as an independent authority that integrated functions previously covered by different bodies [1]. Stipulations concerning nuclear emergency preparedness as well as response were revised based on experiences from the accident. One of the most important missions in response to any nuclear accident is to assess radiation exposure doses of the affected population. However, this is a very challenging task in Japan where hundreds of thousands of people may live within the Urgent Protective action planning Zone (UPZ). Thyroid exposure due to intake of radioiodine is particularly difficult to assess due to its short-lived nature. On the other hand, the International Atomic Energy Agency (IAEA) has recently provided generic criteria (GC) for protection actions, one of which is related to thyroid exposure (i.e., 50 mSv in the first 7 days) [2]. In compliance with the GC, the Basic Plan for Emergency Preparedness of Japan has stated that assessment of internal exposure by inhalation of radioiodine should be performed within one week. This paper presents a proposal to implement reliable and robust monitoring of large populations to assess their thyroid internal doses not only for medical triage purposes, but also for future epidemiological studies concerning health effects of low-dose radiation.

2 The proposed method

The proposed method consists of three surveys: a screening survey, a detailed survey and an additional survey (Figure 1). The screening survey is to be performed within the first week after a nuclear accident with the main purpose of identifying persons who received significant exposure and performing medical triage. The screening survey will be performed by first responders using a simple hand-held device (e.g., a NaI(Tl) survey meter). This type of measurement was conducted in the FDNPP accident; however, it was delayed by more than two weeks due to the chaotic circumstances [3]. Within the following

* Corresponding author: kurihara.osamu@qst.go.jp

three weeks, a more detailed survey using a sophisticated device (e.g., a spectrometer) is to be performed. This survey will be performed by radiation measurement experts at shelters with low background levels after evacuation of residents living near the site of the accident is completed. The screening survey and the detailed survey are direct thyroid measurements in which a probe is placed on (or near) the front surface of the bottom of the neck along with appropriate background subtraction. A final, additional survey is to be performed using whole-body counters (WBCs) after iodine-131 (^{131}I), the largest contributor to the thyroid dose, cannot be detected anymore. In the FDNPP accident, the period within which ^{131}I could be sufficiently detected was only about one month. The target radionuclides of the additional survey are radiocesium, mainly ^{134}Cs and ^{137}Cs. Several mobile WBC units will be available for this survey because many units have been introduced following the FDNPP accident. Both the screening survey and the additional survey will overlap with the detailed survey, which has the highest accuracy, in order to ensure the measurement accuracy and accurately estimate the intake ratio of ^{131}I to $^{134}Cs/^{137}Cs$ in individuals, respectively. The target numbers of subjects for the detailed and additional surveys are currently set at 10,000 and 30,000, respectively.

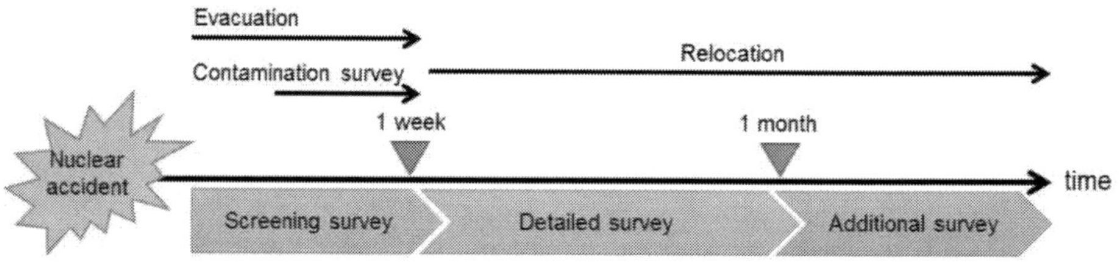

Fig. 1. The proposed method for monitoring populations to assess thyroid internal exposure in a nuclear accident.

3 Discussion

The proposed method intends to effectively use all human data obtained for assessing thyroid internal doses of the affected population. This method is considered robust because it utilizes different techniques suited for each phase. However, the most important objective is to coordinate monitoring of large populations during their evacuation by administrative bodies. Regarding the technical aspects of the three-phase system, the authors have performed various activities, such as training personnel, developing a new device for use in the detailed survey, and considering arrangements for the surveys at a shelter. These activities will be presented in the upcoming workshop.

This work is partly supported by the Nuclear Regulation Authority (NRA) of Japan.

References

1. Nuclear Regulation Authority of Japan, http://www.nsr.go.jp/english/index.html (Accessed on 9 March 2018)
2. International Atomic Energy Agency, *IAEA Safety Standards*, General Safety Guide No. **GSG-2** (IAEA, 2011)
3. E. Kim, O. Kurihara, T. Suzuki, M. Matsumoto, K. Fukutsu, Y. Yamada, N. Sugiura, M. Akashi, *Proc. of the first NIRS symposium on reconstruction of early internal dose in the TEPCO Fukushima Daiichi nuclear power station accident*, **NIRS-M-252**, 59-66 (National Institute of Radiological Sciences, 2012)

Development of a function calculating internal dose coefficients based on ICRP 2007 Recommendations

Kentaro Manabe[1,*], *Kaoru* Sato[1], and *Fumiaki* Takahashi[1]

[1]Nuclear Safety Research Center, Japan Atomic Energy Agency, 2-4 Shirakata, Tokai, Ibaraki, Japan

1 Introduction

Current Japanese regulatory standards for radiation protection against internal exposures are established based upon committed effective dose coefficients in accordance with 1990 Recommendations of the International Commission on Radiological Protection (ICRP) [1]. There are three kinds of standards: derived air concentrations for work places, and concentration limits for exhaust and drainage from facilities. Therefore, dose coefficients for workers and age dependent dose coefficients for members of the public based upon 2007 Recommendations of ICRP [2] are necessary to fit the standards to the 2007 Recommendations.

ICRP have been publishing a series of publications about occupational intakes of radionuclides (OIR) [3-5] and a data base recording new dose coefficients for workers, OIR Data Viewer [4]. As of April 2018, dose coefficients about 28 elements have been released. However, there are some nuclides which are listed in the Japanese standards, and which are not included in neither the publications nor the data base. In addition, a domestic technical basis of internal dose estimation is important to consider various conditions: characteristics of Japanese, specifics of radiation accidents, for example. On these backgrounds, the authority decided to develop a computation code for internal dosimetry based on 2007 Recommendations of ICRP as a public offering four-year project from 2017, and then Japan Atomic Energy Agency (JAEA) is in charge of this project. The code is planned to have two major functions: calculation of dose coefficients and estimation of intake amount from monitoring data. This paper describes the development of the former function.

2 Overview of the function

A programming language, Java, is adopted to develop the code because Java applications are executable on multiple operating systems. The function calculating committed effective dose coefficients, e (Sv Bq^{-1}), uses a dosimetric methodology summarized in OIR part 1 [3]. Figure 1 shows a flow chart of the function. This function calculates e from basic models and data: nuclear decay data [6], specific absorbed fractions (SAFs) [7], biokinetic models [3-5], and radiation weighting factors and tissue weighting factors [2, 3], as shown in Figure 1. In calculating radiation weighted S values, S_w (Sv), Piecewise Cubic Hermite

* Corresponding author: manabe.kentaro@jaea.go.jp

Interpolating Polynomial (PHICP) [8] is used to interpolate the SAFs by radiation energies. In calculating activities and committed equivalent doses, Livermore Solver for Ordinary Differential Equations (LSODE) [9] is applied to solve numerically ordinary differential equations about time dependent changes in activities and equivalent dose rates. To unify the programming language used in the code, the solver packages of PCHIP and LSODE written in Fortran were translated to Java by JAEA; the names of the translated packages are J-PCHIP and J-LSODE. Element specific systemic data are prepared about 14 elements described in OIR part 2.

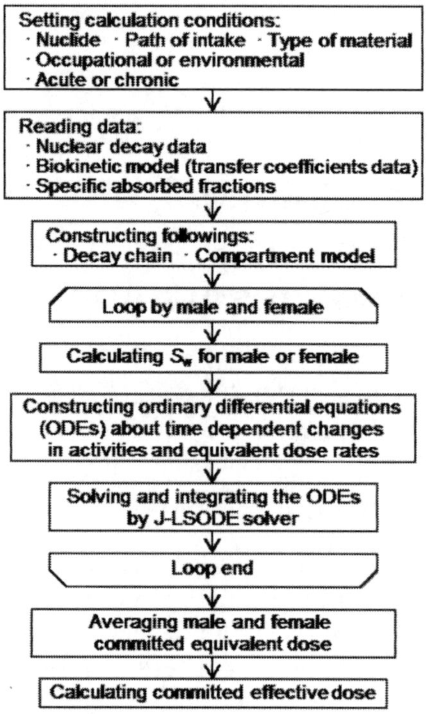

Figure 1. Flow chart of a function calculating committed effective dose coefficients.

3 Quality assurance of the function

Calculation results of *e* by the function were compared to the values of *e* recorded in OIR Data Viewer ver. 2.17.10.17 [4]. The number of intake patterns was 454 for 101 kinds of radionuclides. In 426 cases, the values by the function were consistent with the reference values in two digits which is the number of digits of the values recorded in the viewer. In the residual 28 cases, the differences were only 1 in the second digit. These differences can be explained by rounding errors. Consequently, the quality of the function was assured for the 14 elements. However, the nuclides which emit alpha particle, or decay by neutron fission fragments are not included in the 14 elements. Therefore, an additional quality assurance plan will be necessary in the future.

This study was funded by Nuclear Regulation Authority of Japan.

References

1. ICRP Publication 60 (1991).
2. ICRP Publication 103 (2007).
3. ICRP Publication 130 (2015).
4. ICRP Publication 134 (2016).
5. ICRP Publication 137 (2017).
6. ICRP Publication 107 (2008).
7. ICRP Publication 133 (2016).
8. Netlib Repository at UTK and ORNL. slatec/pchip, http://www.netlib.org/slatec/pchip/.
9. A.C. Hindmarsh, Description and Use of LSODE, the Livermore Solver for Ordinary Differential Equations, UCRL-ID-113855 (1993).

Preliminary outcomes of the ICIDOSE exercise and impact of the new models for occupational intakes.

Augusto Giussani[1,*], *Carlo-Maria* Castellani[2], *Gareth* Roberts[3], *Andor* Andrasi[4], and *Tamas* Pazmandi[4]

[1]BfS - German Federal Office for Radiation Protection, 85764 Oberschleißheim, Germany
[2]ENEA, Radiation Protection Institute, 40129 Bologna, Italy
[3]Nuvia Ltd., Approved Dosimetry Services, Harwell OX11 0RL, United Kingdom
[4]Hungarian Academy of Sciences Centre for Energy Research, Radiation Protection Department, 1525 Budapest POB 49, Hungary

1 ICIDOSE 2017

In 2017 Working Group 7 of the European Radiation Dosimetry Group (EURADOS WG7) has organized an intercomparison action for internal dose assessment for occupational exposures. Main aim of the intercomparison action was to test the practical applicability of the "Technical Recommendations for Monitoring Individuals for Occupational Intakes of Radionuclides" ("TECHREC Recommendations") [1]

Among the four proposed cases, Case 1 (an artificial case of accidental inhalation of ^{60}Co) was focussed on the impact of the changes introduced by the most recent ICRP recommendations [2], including the new report series on Occupational Intakes of Radionuclides (OIR) [3-5]. Additionally, the analysis of the results submitted for Case 4 (a complex real case of accidental inhalation of ^{241}Am) provided useful feedbacks on the revised biokinetic models of ICRP.

2 The new ICRP recommendations and their impact on incorporation monitoring

In its 2007 recommendations ICRP introduced several changes affecting the estimation of the effective dose, among others new values for the radiation and for the tissue weighting factors, new definitions for the "remainder" tissue and for the calculation of the related equivalent dose, adoption of separate male and female reference computational phantoms and a new algorithm for the calculation of the effective dose. In addition to that, a new model for the Human Alimentary Tract (HAT), a revised model of the Human Respiratory Tract Model (HRTM) and updated systemic models for many radionuclides have been presented (3-6). All these changes affect not only the estimation of the effective dose, but also the calculation of the retention and excretion functions used to interpret the incorporation measurements.

* Corresponding author: agiussani@bfs.de

3 ICIDOSE Cases

Case 1 was an artificial case purposely conceived to show that the two different approaches (analysis with current definitions and data vs. those adopted in the OIR reports) not only affect the final estimates of intake and dose, but also the procedure recommended by the TECHREC document for those estimates, requiring different levels of effort and expertise.

On the other side, the analysis of the results provided for Case 4 showed that the revised HRTM model and the OIR systemic model for americium radionuclides can successfully describe the retention data as measured in the real individuals. Following the procedure described in the IDEAS Guidelines (Version 2) for the Estimation of Committed Doses from Incorporation Monitoring Data [7], a very good fit of the data can be achieved with only slight changes of the HRTM americium specific parameters given in the draft of Part 4 of the OIR Report Series [8].

Finally, this analysis also enabled to check the applicability of the "dose per content" concept presented in the OIR Reports as an alternative method to estimate committed doses without the need to assess the related intake.

References

1. G. Etherington, A. Giussani, M.A. Lopez, P. Bérard, E. Blanchardon, B. Breustedt, C.M. Castellani, D. Franck, J. Marsh, D. Nosske, C. Challeton-de Vathaire. Technical Recommendations for Monitoring Individuals for Occupational Intakes of Radionuclides. EC Radiation Protection series (in press, 2018).
2. International Commission on Radiological Protection. The 2007 Recommendations of the International Commission on Radiological Protection. ICRP Publication 103. Ann. ICRP 37(2-4) (2007).
3. International Commission on Radiological Protection. Occupational Intakes of Radionuclides: Part 1. ICRP Publication 130. Ann. ICRP 44(2) (2015).
4. International Commission on Radiological Protection. Occupational Intakes of Radionuclides: Part 2. ICRP Publication 134. Ann. ICRP 45(3-4) (2016).
5. International Commission on Radiological Protection. Occupational Intakes of Radionuclides: Part 3. ICRP Publication 137. Ann. ICRP 46(3-4) (2017).
6. International Commission on Radiological Protection. Human Alimentray Tract Model for Radiation Protection. ICRP Publication 100. Ann. ICRP 36(1-2) (2006).
7. C.M. Castellani, J.W. Marsh, C. Hurtgen, E. Blanchardon, P. Berard, A. Giussani, M.A. Lopez, EURADOS report 2013-01 (2013) - available at http://www.eurados.org/en/Documents_Publications/Reports_documents
8. International Commission on Radiological Protection. Occupational Intakes of Radionuclides: Part 4. In preparation (Draft for public consultation published in July 2016).

Multi-cellular dosimetry of β+-emitting radionuclides used for cell labeling in the context of cell tracking studies with PET imaging

Manon Jacquemin[1], *Charles* Truillet[2], *Benoit* Jego[2], *David* Broggio[1], *Didier* Franck[3], *Jean-Marc* Bertho[4] *and Aurélie* Desbrée[1]

[1]*IRSN/PSE-SANTE/SDOS/LEDI, Fontenay-aux-Roses*
[2]CEA DRF/I2BM/*SHFJ*/U1023 Service Hospitalier Frederic Joliot, Orsay
[3]*IRSN/PSE-SANTE/SDOS, Fontenay-aux-Roses*
[4]*IRSN/PSE-SANTE/SESANE/LRTOX, Fontenay-aux-Roses*

Introduction

Radiolabeling of cells combined with nuclear medicine imaging provides a potential method for cell trafficking analysis *in vivo*. Besides, the use of β^+-emitting radionuclides is of particular interest given the good intrinsic characteristics offered by PET technology. Generally, a direct labeling approach is applied in which cells are mixed with a radioactive solution and incubated *in vitro* at 37°C to permit the incorporation of the radionuclide intracellularly. Because the procedure implies the use of high amount of radioactivity, various studies have taken interest about the effects of labeling on cell viability and functionality. Though severe cell cytotoxicity was pointed out, *e.g.* reduction of proliferation capacity when labeling with ^{18}F-FDG [1, 2], not clear conclusions has been drawn from these findings. Particularly, comparison and analysis of results are limited because the observed biological effects are often evaluated as a function of added activity rather than absorbed dose delivered to the cell [3, 4, 5]. Accurate dosimetry is required to allow for the establishment of a reliable absorbed dose-effect relationship and therefore to get a comprehensive assessment of the toxicity induced by the labeling procedure. This work aimed to develop a realistic multi-cellular dosimetry and apply the model to the analysis of 3 radionuclides used for labeling in PET: ^{18}F, ^{64}Cu and ^{68}Ga.

Materials and Methods

A program using Python was developed to generate cell coordinates from a set of initial parameters i.e. cell size, cell density, bounding volume size. A cubic cell arrangement was assumed and the calculation approach was based on an analytical formula using pre-determined S factors (absorbed dose per unit time-integrated activity). These S values were calculated by MCNP6 modelling considering radioactivity uniformly distributed in a source cell placed at variable distances from another cell with the nucleus as the target. The mean absorbed dose to the cell nucleus was determined as a function of the cell density and the intra-to-extra cellular activity distribution considering both the contribution of radioactivity located within the cells (self- and cross-dose) and in the culture medium. To verify the contribution of cellular dosimetry in this study, the results were compared with those obtained using conventional dosimetry which is based on the assumption that the radionuclide is distributed homogenously in the global volume. Dose calculations were also done under realistic conditions of labeling of leukocytes with ^{18}F-FDG on the basis of labeling parameters used in few reported studies.

Results

For all radionuclides, the cellular-to-conventional dose ratios can vary significantly with cell density, reaching important values at low density and approaching 1 at high density. They were also influenced by the activity distribution between the cells and the medium, the highest values occurring when all activity was incorporated in the cells. Such trends were in agreement with previous work [6, 7]. A significant underestimation of mean dose to the cell by conventional model was particularly observed for ^{64}Cu up to 15-fold at low density when source is located in the nucleus. This observation is due to the emission of short-range auger electrons, responsible for a major self-dose, that deposit all their energy in the cell nucleus. ^{68}Ga and ^{18}F presented lower discrepancies between cellular and conventional doses considering the emission of larger-range β^+ that enables to counteract a certain degree of heterogeneity in the source distribution by a more homogenous dose deposit. Finally, the comparison between published data highlighted the high disparity of parameters used for labeling of leukocytes with ^{18}F-FDG. Calculation results showed significant deviations of the actual cell dose from the conventional one up to a factor of 5. We also observed that a same injected activity per cell can result in important different absorbed doses.

Conclusions and Perspectives

The development of a realistic multi-cellular dosimetry offered a better understanding of how absorbed dose to cell is affected as a function of key labeling parameters such as uptake ratio of activity within cells and cell density. The need to use cellular model over the conventional one was highlighted in the case of low cell densities and high activity incorporation into cells, especially when labeling with radionuclides emitting low range electrons such as ^{64}Cu. The establishment of the dose-effect relationship for mesenchymal stem cells is under way through functional tests *in vitro* including clonogenic assay, proliferation ability and senescence induction after labeling with ^{18}F-FDG and external irradiation for comparison.

References

[1] C. Botti *et al* (1997). Comparison of three different methods for radiolabelling human activated T lymphocytes. *EJNM*, **24:497–504**.
[2] E. Miñana *et al* (2015). Quantification of the chromosomal radiation damage induced by labelling of leukocytes with [18F]FDG. *Nucl. Med. Biol,* **42:720–723**.
[3] O. Detante *et al* (2009). Intravenous Administration of 99m Tc-HMPAO-Labeled Human Mesenchymal Stem Cells After Stroke : In Vivo Imaging and Biodistribution. *Cell transplantation*, **18:1369–1379.**
[4] F. J. Gildehaus *et al* (2011). Impact of Indium-111 Oxine Labelling on Viability. *Mol Imaging Biol*, **13:1204–1214**.
[5] E. Wolfs *et al* (2017). 18F-FDG Labeling of Mesenchymal Stem Cells and Multipotent Adult Progenitor Cells for PET Imaging: Effects on Ultrastructure and Differentiation Capacity. *J. Nucl. Med,* **447–455**.
[6] M. Faraggi *et al* (1998). Original article Comparison of cellular and conventional dosimetry in assessing self-dose and cross-dose delivered to the cell nucleus by electron emissions, *Eur J Nucl Med*, **25:205–214**.
[7] G. M. Makrigiorgos *et al* (1989). Limitations of Conventional Internal Dosimetry at the Cellular Level, *J. Nucl. Med*, **30:1856–1865**.

Dose assessment for workers involved in an internal contamination accident with Pu at JAEA's Oarai R&D Center

Eunjoo Kim [1,*], *Osamu* Kurihara[1], *Masako* Ohno[1], *Kazuaki* Yajima[1], *Takashi* Nakano[1], *Kotaro* Tani[1], *Nobuhito* Ishigure[1], *Chie* Takada[2], *Takumaro* Momose[2], *Takako* Tominaga[2], *Hideo* Tatsuzaki[2] and *Makoto* Akashi[3]

[1]National Institutes for Quantum and Radiological Science and Technology (QST), National Institute of Radiological Science and Technology (NIRS), 4-9-1 Anagawa, Inage-ku, Chiba-city, Chiba 263-8555, Japan
[2]Nuclear Fuel Cycle Engineering Laboratories (NCL), Japan Atomic Energy Agency (JAEA), 4-33 Muramatsu, Tokai-mura, Naka-gun, Ibaraki 319-1194, Japan
[3]QST

1 Introduction

An internal contamination accident with plutonium (Pu) compounds occurred at a facility of the Oarai Research and Development Center of the Japan Atomic Energy Agency (JAEA) on June 6, 2017. The outlines of this accident are provided elsewhere [1]. On the day following the accident, the National Institues for Quantum and Radiological Sceinces and Technology(QST-NIRS) received the five workers who were involved in the accident and has continued direct and indirect measurements for internal dose assessments as well as treatments using chelating agents (Ca-DTPA, Zn-DTPA). The first dose assessment results of the five workers were reported on July 10, 2017; the highest dose estimate for the workers was found to be within 100–200 mSv as a committed effective dose (CED) [2]. This accident was the first internal contamination accident in which a decorporation therapy to remove inhaled Pu outside the body was performed in Japan. The decorporation therapy continued more than one month for three of the five workers. This paper describes highlights regarding the first dose assessment and some future developments, in particular for evaluating the dose reduction due to decorporation therapy.

2. The first dose assessment

The aim of the first dose assessment was to determine the workers' internal doses in a manner that would not result in an excessive overestimate, based on a reasonable interpretation of the measured results of faeces, urine samples, and the lungs. Regarding the

*Corresponding author; email: kim.eunjoo@qst.go.jp

lung counting with high-purity germanium (HPGe) detector arrays, americium-241 (^{241}Am) was detected from two of the five workers. The effective size of the inhaled Pu aerosols was estimated from the ^{241}Am activity ratio of the early faecal excretion (collected for the first 5 days after intake) to the residual lung deposition (Figure 1) [3]. The size determined was 5 μm in Activity Median Aerodynamic Diameter (AMAD) for the worker with the highest dose (Worker A) and 1 μm in AMAD for the other four workers. These sizes have been designated as default values for occupational exposure [4]. The absorption type in the respiratory tract was estimated from the urine samples collected prior to the first decorporation therapy using Ca-DTPA (calcium diethylentriamene pentaacetate) that was performed about half a dayafter the workers' intake at the JAEA-NCL. In a comparison with the predicted urinary excretion of Pu and ^{241}Am using the intake amounts calculated from the early faecal excretion, the results suggested a mixture of Type M and Type S materials (Figure 2). Type M (with higher dose coefficients for Pu) was thus selected to avoid underestimations of the doses.

3. Future development of the dose assessment

To evaluate dose reductions due to decorporation therapy, it is necessary to assess more accurate intake amounts as well as to develop a sophisticated model in which the impact of Ca/Zn-DTPA on the biokinetic behavior of Pu can be identified [5]. As described, the aerosol size of 5 μm in AMAD was determined for Worker A; however, larger aerosols are expected from the results (Figure 1). The early faecal excretion of ^{241}Am for Worker A was ~700 Bq, so that the predicted ^{241}Am lung deposition was ~100 Bq assuming 5 μm, which is much larger than the worker's lung counting result (20 Bq). Figures 3 and 4 are the preliminary calculations to explain the dose reduction for Worker A. Figure 3 compares the observed and predicted Pu urinary excretion (Type M, Type S, and Mix) for Worker A, indicating that the difference between the observed and predicted values is related to the dose reduction. Here the predicted values were obtained assuming the aerosol size of 15 μm, and those for Mix were obtained by adjusting the ratio of Types M and S to match the observed urinary excretion prior to the first decorporation therapy. Figure 4 shows the predicted residual Pu activity of blood, liver, bone and lungs for Mix without the effects of Ca/Zn-DTPA and indicates that the efficacy of Ca/Zn-DTPA at later days is very limited. However, a temporal increase in urinary excretion was observed in the urinary data obtained several months after the workers' intake, suggesting that the absorption parameters in the respiratory tract should be modified. Further analyses will be presented at the conference.

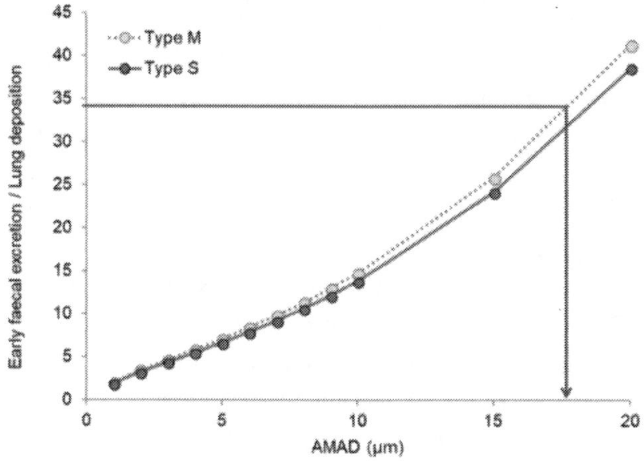

Figure 1. Estimation of the effective size of inhaled aerosols.

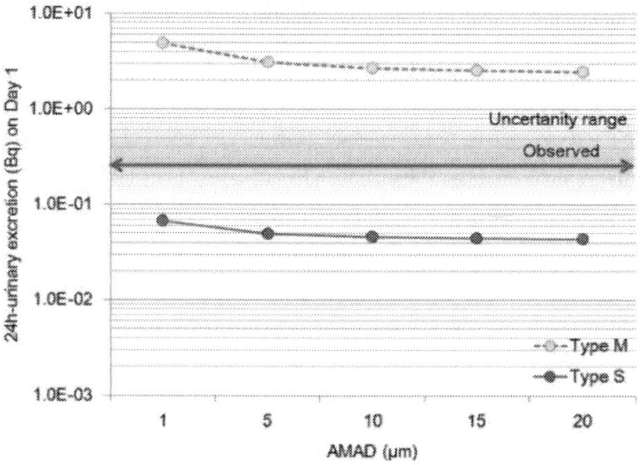

Figure 2. Estimation of the absorption type.

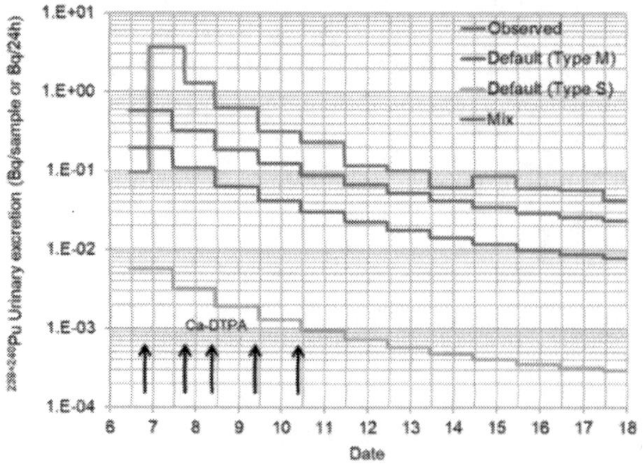

Figure 3. Comparison of the observed and predicted Pu urinary excretions.

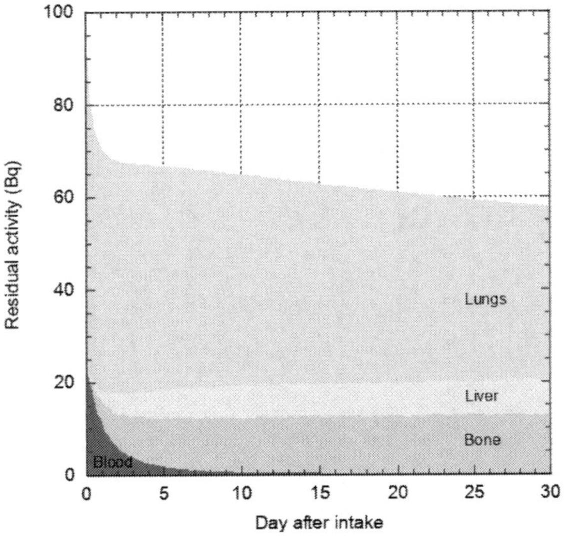

Figure 4. Predicted Pu residual activities in various parts of the body

References

1. Tatsuzaki et al. Radiat. Prot. Dosim (accepted)., Eur. Phys. J. E **14**, 7 (2004)
2. Japan Atomic Energy Agency, Report on contamination at the Plutonium Fuel Research Facility in the Oarai Research and Development Center (The third report) (in Japanese), Japan Atomic Energy Agency, 29 September (2017). https://www.jaea.go.jp/02/press2017/p17092902/ (accessed 2018-5-14)
3. IDEAS General Guidelines. General Guidelines for the Estimation of Committed Effective Dose from Incorporation Monitoring Data. FZKA 7243 (2006). http://www.bologna.enea.it/Rapporti/General%20Guidelines-FZK-report.pdf (accessed 2018-5-14)
4. International Commission on Radiological Protection. Dose Coefficients for Intakes of Radionuclides by Workers. ICRP Publication 68 Ann. ICRP 24 (4), Pergamon Press (1994)
5. Tani et al., This conference

Localized Instantaneous Dose Rates from Inhaled Particles of ^{239}Pu

Lucas Hetrick[1,*], *John* Klumpp[1], *Deepesh* Poudel[1]

[1]Radiation Protection Division, Los Alamos National Laboratory, Los Alamos, NM

1 Introduction

Inhalation of plutonium, specifically ^{239}Pu, is a topic of importance at nuclear weapons laboratories such as Los Alamos National Laboratory. The stochastic risk resulting from an inhalation event is represented by the 50-year CED expected to result from the intake, which is calculated by multiplying the activity which was inhaled by the corresponding ICRP-supplied dose coefficients. However, estimating the localized dose to lung tissue from inhaled ^{239}Pu is a more challenging task. Inhaled particles are known to deposit heterogeneously throughout the lung with particles deposited in the upper airway being swept away until they are swallowed and incorporated into the gastrointestinal tract, while particles in the lower airway tend to be less mobile. Particles that remain in the airway for extended periods of time may result in significant dose to the respiratory tract. However, because the particles are deposited heterogeneously, it is inaccurate to calculate the dose to the lung as if the particles were deposited homogenously, or to average the dose over the entire lung volume. This is especially true for alpha-emitters such as ^{239}Pu, where the dose from a given particle is absorbed locally.

It is thought that the carcinogenic effects from a hot particle are less significant than that from a uniform exposure [1, 2]; but literature says very little about the deterministic effects of hot particles in the lungs. For example, the National Council on Radiation Protection and Measurements (NCRP) has published limits for hot particle exposure to the lungs [3]. However, the effects of highly localized dose such as from an alpha-emitting "hot particle" are not discussed. A recent study [4] published dose rates for alpha-, beta- and gamma-emitting hot particles in the lungs of rabbits. However, the resolution of this study was not sufficient to capture the local dose distribution from alpha emitters. It also does not consider that the dose distribution from alpha emitters will be different in different regions of the respiratory tract, each of which has substantially different dimensions.

This work calculates the localized dose to lung tissues from a single particle of ^{239}Pu deposited in various regions of the respiratory tract. These doses might be added and combined to predict deterministic effects in a situation in which an aerosol is inhaled, and thus to set better limits for hot particle exposures to the lungs.

* Corresponding Author: lhetrick@lanl.gov

2 Materials and Methods

Localized dose calculations were performed using version 6.2 of the Monte Carlo N-Particle radiation transport code [5] with a simplified model of an adult human airway. The model was composed of a cylinder of air surrounded by a concentric cylinder of tissue. Calculations were performed using the dimensions for regions of the lower airway (BB, bb, and AI) of an adult male as specified by ICRP Publication 66 [6]. In each case, a particle was embedded half-way into the tissue, such that one hemisphere was entirely embedded in lung tissue, while the other hemisphere was entirely exposed to the air pathway.

The dose rate from the embedded ^{239}Pu hot particle was calculated at three areas of interest for hot particles ranging from 1-100 μm in increments of 10 μm. 1, 10, and 100 nm particles were also simulated. The first area of interest was the tissue that surrounded the hot particle. The second area of interest was located at the boundary of the airway and the tissue surrounding the particle. The third area of interest was the tissue across the airway from the hot particle. All the material compositions used in the models were taken from the Compendium of Material Composition Data for Radiation Transport Modeling [7].

3 Discussion

It is of interest to examine what, if any, insights this study may provide into a single particle's contribution to the induction of deterministic effects. The alpha particles from ^{239}Pu have a short range in tissue, and an even shorter range in plutonium. Although the volume irradiated by a single plutonium particle is not sufficient to result in observable deterministic effects, the cumulative volume irradiated by many plutonium particles can cause cell morbidity and mortality. While the irradiated volume differs depending on particle size, the dose rate is such that the irradiated volume would be killed in a matter of days. It is plausible that most of the alpha energy emitted by a low-solubility particle would be 'wasted' on dead tissue, which would act like shielding against healthy tissue, while mobile particles might have a greater ability to induce deterministic effects than stationary particles. This implies that particles of ^{238}Pu, which are known to break up and transport themselves around the airway due to their decay energy [8], may be more harmful per unit of activity than ^{239}Pu. This hypothesis could partially explain the anomaly observed in a canine study that experienced a higher than expected occurrence of pulmonary fibrosis from inhaled $^{238}PuO_2$ [9].

References

1. Coggle J, Lambert B, Moores S. Env health persp. 70; 261; 1986.
2. Lang S, Servomaa K, Kosma VM, Rytömaa T. Env health persp. 103.10:920; 1995.
3. National Council on Radiation Protection and Measurements. Report No 130 (1999)
4. Caffrey E, Johansen M, Caffrey J, Higley K. Health Physics 112.6: 526-532; 2017
5. Werner CJ, et al. Los Alamos National Laboratory. Report LA-UR-18-20808 (2018)
6. ICRP Publication 66 (1994)
7. McConn RJ, Gesh CJ, Pagh RT, Rucker RA, Williams III R. PNNL. (2011)
8. Mewhinney JA, Diel JH. Health Phys 45: 39–60; 1983.
9. Muggenburg BA, Guilmette RA, Mewhinney JA, Gillett NA, Mauderly JL, Griffith WC, Diel JH, Scott BR, Hahn FF, Boecker BB. Radiat Res 145:361–381; 1996.

ICARE software for calculation of dose coefficients and retained/excreted fractions of intake

Eric Blanchardon[1,*], *Estelle* Davesne[1], *Sandra* Bohand[2], and *Pierre* Laroche[2]

[1]IRSN, Fontenay-aux-Roses, France
[2]ORANO, Paris, France

The assessment of internal exposure for workers is usually based on ambient activity measured at the workplace or individual activity from bioassay interpreted with dose coefficients and time-dependent retained and excreted fractions of intake established by the International Commission on Radiological Protection (ICRP) from reference biokinetic and dosimetric models.

To gain flexibility in the implemented models and enable consideration of non-standard situations of exposure such as wound contamination, clinical trials of radiopharmaceuticals, exposure to radon decay products with specific aerosol parameters or intake followed by decorporation treatment, a new software module, called "Internal dose CAlculation for Radionuclide Exposure (ICARE)", was developed for expertise and research. It provides the dose coefficients and retained/excreted fractions used by OPSCI software [1] (IRSN-ORANO) which performs occupational dose calculation from bioassay measurement, evaluates the associated uncertainty and supports design of monitoring programs of internal exposure.

ICARE will be presented as software developed with the Interactive Data Language (IDL, Harris) to implement biokinetic models of the ICRP 60 [2] series of publications by the method described by Leggett et al. [3]. It provides retained fractions of intake in body regions, daily urinary and fecal excretions after occupational exposure by inhalation, ingestion or wound [4]. S coefficient values are calculated from emission spectra of ICRP [5] convoluted with specific absorbed fractions published by Cristy and Eckerman [6] for photons and by ICRP [7, 8] for electrons and alpha particles. S coefficients are combined with numbers of nuclear transformations to provide absorbed, equivalent and effective doses per unit intake.

For validation, calculation results were extensively compared with ICRP equivalent and effective dose coefficients [9] and with bioassay quantities predicted by DCAL [10] and IMBA [11] software. Examples for specific exposure situations will be presented. Further development is planned to include age-dependent models and revised ICRP models implemented in the Occupational Intakes of Radionuclides series of publications [12] with the objective of contributing to the quality assurance of the on-going update of ICRP dose coefficients for internal exposure.

References

[1] E. Davesne, M. Vincent, E. Blanchardon, M. Rennesson, P. Laroche, D. Franck. IAEA(ed.) Vienna (2014) Available at: http://www-ns.iaea.org/tech-areas/communication-networks/orpnet/documents/CN223_Book%20of%20contributed%20papers.pdf

[2] ICRP Publication 60. Ann. ICRP **21** (1-3) (1991)

* Corresponding author: eric.blanchardon@irsn.fr

[3] R.W. Leggett, K.F. Eckerman, L.R. Williams, Health Phys. **64**, 260 (1993)

[4] National Council on Radiation Protection and Measurements. Bethesda, MD: NCRP; Report No. 156 (2006)

[5] ICRP Publication 38. Ann. ICRP **11-13** (1983)

[6] M. Cristy, K.F. Eckerman. ORNL/TM-8381/V1-7, Oak Ridge National Laboratory, Oak Ridge, TN (1987)

[7] ICRP Publication 30 (Part 1). Ann. ICRP **2** (3-4) (1979)

[8] ICRP Publication 66. Ann. ICRP **24** (1-3) (1994)

[9] ICRP. Available at http://www.icrp.org/docs/ICRPDOSE_setup.exe.

[10] K.F. Eckerman, R.W. Leggett, M. Cristy, C.B. Nelson, J.C. Ryman, A.L. Sjoreen, R.C. Ward. Oak Ridge, TN: Oak Ridge National Laboratory; Report ORNL/TM-2001/190 (2006)

[11] A. Birchall, M. Puncher, J.W. Marsh, K. Davis, M.R. Bailey, N.S. Jarvis, A.D. Peach, M.-D. Dorrian, A.C. James, Radiat. Prot. Dosim. **125**, 194 (2007)

[12] ICRP Publication 130. Ann. ICRP **44**(2) (2015)

BIO Web of Conferences **14**, 03017 (2019) https://doi.org/10.1051/bioconf/20191403017
HEIR 2018

Evaluation of the activity of a hot particle incorporated by a worker of nuclear power plant

Mathieu Sailly[1], *Isabelle* Le-Couteulx[2], *Laetitia* Rollin[1], *Jean-François* Gehanno[1] and *Ismérie* Bocquet[2]

[1] CHU Rouen, 76000 Rouen, France
[2] EDF CNPE de Paluel, BP 48, 76450 Cany Barville, France

1 Introduction

Nuclear power plants workers can be exposed to internal contamination via hot particles (strongly radioactive insoluble particles with a 10µm to 1mm wide diameter). At present, no theoretical biokinetic model describing the evolution of an internal contamination by these particles has been designed by the ICRP [1][2].In addition, the procedure that allows the evaluation of an internal contamination by diffuse radioactive substances cannot be applied. The objective of this work was to describe a new method to estimate the activity of a hot particle in the body.

2 Material and method

Several whole body measurements were realized using an anthropomorphic ghost called "Igor" into which a radioactive source of known activity was inserted. Igor is constituted with elementary bricks allowing to feign the various locations of a hot particle within the abdomen. These various measures allowed to establish a corrective factor in the activity measured for every location of the source.
To determine the position of the hot particle in the abdomen, two whole body measures were successively realized (Face then Back) at the contaminated worker.
The difference between these two examinations allows to estimate the particle's position. The corrective factor corresponding to the estimated position is then applied to the moderate activity.
This method was then used in a real case of internal contamination, in parallel feces analysis were realized. The activity obtained by our method was compared with the activity obtained by the feces analysis.

3 Results

https://doi.org/10.1051/bioconf/20191403017

In total, 24 measures were realized to consider various locations of the hot particle in the abdomen, the weight of the subject (configuration 70-90kg) and the type of radionuclides involved. The corrective factor (FC) was 1.20-1.47 (source in front position) and 0.48-0.56 (source in posterior position) in configuration 70 kg and of 0.70-1.05 and 0.48-0.56 in configuration 90 kg.

The application of this method in a real case of internal contamination allowed to estimate the activity of a hot particle between 76,800 and 90,250Bq. The activity contained in the abdomen according to the fecal examinations was 82,523Bq.

4 Conclusion

This new method allows to estimate more quickly the real activity of a hot particle and could be a complementary tool to feces analysis as collecting is sometimes troublesome.

References

1. F. Paquet, M.R. Bailey, R.W. Leggett, J. Lipsztein, T.P. Fell, T. Smith, D. Nosske, K.F. Eckerman, V. Berkovski, E. Ansoborlo, and A. Giussani, W.E. Bolch, J.D. Harrison, 'ICRP, 2016. 'Occupational Intakes of Radionuclides: Part 2. ICRP Publication 134. Ann. ICRP 45(3/4), 1–352'.

2. F. Paquet, G. Etherington, M.R. Bailey, R.W. Leggett, J. Lipsztein, and W. Bolch, K.F. Eckerman, J.D. Harrison, 'ICRP, 2015. Occupational Intakes of Radionuclides: Part 1. ICRP Publication 130. Ann. ICRP 44(2).'

Accounting for environmental intakes when assessing occupational exposures to uranium

Derek Bingham[1,*] and *Rupert* Cockerill[1]

[1]Dosimetry Service, AWE, Reading, RG7 4PR, UK

Identifying occupational exposure to uranium is complicated by the presence of varying amounts of natural uranium found in urine samples due to environmental (dietary) intakes [1]. At AWE, when the total uranium activity in urine in a sample is reported above the investigation level (2.5 $mBqd^{-1}$), then several lines of enquiry are made to determine whether this is of occupational or environmental origin. They include: obtaining a further sample; obtaining information on the person's diet and work history; investigating uranium isotope ratios using alpha and mass spectrometry.

An increase in the number of urine samples with total uranium activity above investigation levels has been observed at AWE in recent years. This appears to be related to the provision on site of a new source of bottled drinking water with a relatively high content of natural uranium. However, it is not always possible to determine with confidence the origin of an elevated uranium in urine result. There can be significant consequences, in terms of the assessed dose, from incorrectly attributing the origin of the exposure.

References

1. [EURADOS 2013] Castellani, C. M., Marsh, J. W., Hurtgen, C., Blanchardon, E., Bérard, P., Giussani, A. and Lopez, M. A. IDEAS Guidelines (Version 2) for the Estimation of Committed Doses from Incorporation Monitoring Data. EURADOS Report 2013-01 ISBN 978-3-943701-03-6 (2013).

   http://www.eurados.org/~/media/Files/Eurados/documents/EURADOS%20Report%20 2013-01%20online%20version.pdf?la=en

* Corresponding author: Derek.Bingham@awe.co.uk

BIO Web of Conferences **14**, 03019 (2019) https://doi.org/10.1051/bioconf/20191403019
HEIR 2018

^{223}Ra dichloride incidental inhalation: recommendations to estimate the committed effective dose

Dominique Saurat[1,*], *Olivier* Aupée[2], *Éric* Gontier[3], *Denis* Métivier[4], *Alain* Cazoulat[1], and *Yannick* Lecompte[1]

[1]Service de protection radiologique des armées, 1bis rue du Lieutenant Raoul Batany, CS500 57, 92141 Clamart Cedex, France
[2]Hôpital d'instruction des armées Percy, Service de pharmacie hospitalière, 101 avenue Henri Barbusse
BP 406, 92141 Clamart Cedex, France
[3]Services de médecine nucléaire, Institut inter-régional de cancérologie, Centre Jean Bernard 72000 Le Mans, et Centre d'Imagerie Scintigraphique d'Eure et Loir, 28300 Le Coudray, France.
[4]Hôpital d'instruction des armées Sainte Anne, 2 boulevard Ste Anne, BP 600, 83800 Toulon, France

1 Introduction

^{223}Ra dichloride - Xofigo® (Bayer AG, Germany) - is a therapeutic alpha particle-emitting pharmaceutical used in nuclear medicine for patients with metastatic castration-resistant prostate cancer (mCRPC). The radiopharmaceutical is formulated as a ready-to-use solution (unsealed source) and is administered to patients as an intravenous injection [1-2]. As it is an alpha-emitter, internal contamination is feared. If inhalation contamination is suspected, an adequate special monitoring programme needs to be provided to estimate the committed effective dose. ^{223}Ra internal contamination monitoring can be performed by two methods: *in vivo* measurements and *in vitro* measurements (urine of faeces). The purpose of our study was to determine the most appropriate method for individual monitoring of nuclear medicine staff who could have inhaled ^{223}Ra and propose recommendations for committed effective dose assessment.

2 Materials and methods

First, minimum detectable activities (MDA) and scattering factors (SF) of these methods were estimated according, respectively, to the French working group number 5 (GTN5) and EURADOS Guidelines [3-4]. Concerning *in vitro* analysis, due to the ^{223}Ra short half-life (11.4 d), an adjusted MDA was calculated at the end date of the 24-h sampling to consider transport and sample pre-treatment times (~48h for urine and ~96h for faeces). *In vivo* measurements were obtained by a bed-type whole-body counter equipped with two coaxial p-type high purity germanium (HPGe) detectors. The counting time was set to 45 min. *In vitro* measurements were obtained by direct measurement using a gamma spectrometer equipped with one coaxial n-type HPGe detector on 500-mL aliquot portion of the true 24-h

* Corresponding author: dominique.saurat@hotmail.fr

urine samples or complete 24h-faeces ashed and dissolved in acid. The counting time was set to 10800 s.

Then, the minimum detectable effective dose (MDED), which is the committed effective dose at time *t* after incorporation corresponding to the MDA was calculated by the following equations:

$$MDED(t) = \frac{MDA}{F(t)} * e(50) \quad (1)$$

in which e(50) is the effective dose coefficient (5.7 x 10^{-6} Sv Bq^{-1}) following a unit intake of ^{223}Ra for inhalation (pulmonary absorption parameter "Moderate" and AMAD set to 5 µm by default), F(*t*) is the value of the retention or excretion function at time *t* (in days).

3 Results and discussion

Special monitoring programmes should provide enough data to assess the committed effective dose. That's why a suitable combination of *in vivo* measurements and *in vitro* analyses according to the appropriate biokinetic model would be used.

Figure 1 compares the MDED to an effective dose limit of 1 mSv (workers recording level). It appears that whole body counting (WBC) is sensitive enough only the day following incorporation. Although urine samples analysis has a low SF, it should be used only in a case of a major contamination (>15 mSv). Thus, due to its rapidity and its non-invasiveness, WBC (with HPGe detector) should be the first choice to estimate the committed effective dose. However, after 24 h, sufficient sensitivity can only be reached by true 24-h faeces samples analyses (up to 8 days after contamination). Thus, in that case, despite its main drawbacks (excretion fluctuation, staff reluctance, higher SF…), this method should be associated with WBC to estimate the committed effective dose.

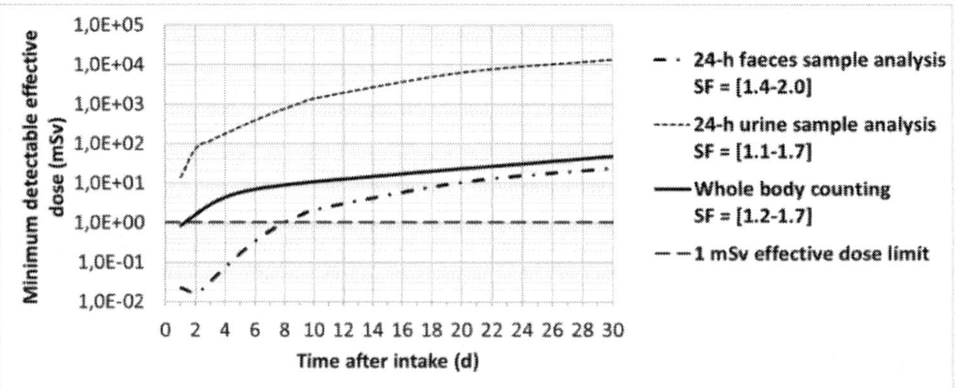

Figure 1: MDED after acute inhalation of ^{223}Ra as aerosol (type M; 5µm AMAD) with whole-body counting, urinary or faecal analyses and SF associated to each method.

References

1. Bayer AG. Annex I - Summary of product characteristics. Xofigo: EPAR - Product Information. (2017).
2. Wadas, T. J. Pandya, D. N. Solingapuram Sai, K. K. and Mintz, A. AJR Am J Roentgenol. **203**(2), 253-260 (2014).
3. Groupe de travail de normalisation n°5 (GTN5) du Centre Technique d'Homologation de l'Instrumentation de Radioprotection (CTHIR) (1994).

4. Castellani, C. M. Marsh, J. W. Hurtgen, C. Blanchardon, E. Berard, P. Giussani, A. Lopez, M-A. EURADOS Report 2013-011. (2013)

BIO Web of Conferences **14**, 03020 (2019) https://doi.org/10.1051/bioconf/20191403020
HEIR 2018

Current radiation risks from occupational exposure in Russian nuclear industry

Viktor Ivanov*, *Alexander* Korelo, *Sergey* Chekin, and *Konstantin* Tumanov

A. Tsyb Medical Radiological Research Center, Obninsk, Russia

For monitoring risks of occupational radiation exposure to workers of Russian nuclear industry the system ARMIR is used. The system was developed at the State Corporation ROSATOM in cooperation with the Russian Scientific Commission on Radiological Protection and the National Radiation Epidemiological Registry [1-3].

The ARMIR system is designed on the basis of principles and methods for calculation of radiation risk recommended by ICRP [4]. During the creation of the system algorithms of calculation of different radiation risk measures from multiple radiation exposure, specialized software and database were developed. Justification of adaptability of mathematical models based on follow-up data on A-bomb survivors in Japan to Russian conditions was carried out.

The system ARMIR was first used in the "Mayak Production Accociation" in 2006. At present the system is used in more than 100 radiation and nuclear hazardous sites of the ROSATOM.

ARMIR databases consist all information necessary for calculating radiation risks, that is employee's ID number, gender, date of birth, as well as annual radiation doses received by a regulatory monitored worker being registered on a database maintaining records of the doses received by each individual dosimeter wearer. More than 60 thousand personal records are collected annually (Table 1).

Table 1. Basic characteristic of occupational radiation exposure in main divisions of the Rosatom.

Division	Number of workers	Average age, years	Average annual dose, mSv	Average cumulative dose, mSv	Average length of work, years
Electric power division	27427	42.0	1.77	51.2	12.4
Fuel division	7637	43.7	1.20	19.7	11.1
Nuclear weapons complex	5131	38.5	3.39	30.8	5.8
Mining division	13959	42.9	1.82	31.4	11.9
Nuclear and radiation safety complex	3334	45.1	1.05	31.0	14.1
Innovation management unit	4578	48.3	2.17	41.2	14.9
ROSATOM	62349	42.8	1.83	39.3	11.9

* Corresponding author: nrer@obninsk.com

During the recent years average risk value does not exceed 0.00008 (Table 2).

Table 2. Results of monitoring radiation risks in main ROSATOM divisions.

Division	Average individual risk	Average Safety Index of Potential Exposure, %
Electric power division	1.2E-04	67.2
Fuel division	3.1E-05	76.5
Nuclear weapons complex	5.0E-05	65.0
Mining division	2.0E-05	48.0
Nuclear and radiation safety complex	6.8E-05	76.9
Innovation managment unit	1.0E-04	67.2
ROSATOM	7.9E-05	67.0

Data accumulated in the ARMIR are used for calculation of the Safety Index of Potential Exposure, SIPE. The index takes into account limitation of annual generic risk value, $< 2 \cdot 10^{-4}$. This allows monitoring radiation protection at ROSATOM facilities with regard to sources of potential exposure. The level of radiation protection at a nuclear facility in relation to sources of potential radiation exposure is considered as optimized if the SIPE is above 50% and does not reduce with time. Average safety index of potential radiation exposure in ROSATOM achieves 67%.

Information on existing radiation risks in nuclear industry is brought to the notice of professional association and interested public via ROSATOM Annual Public Report and industry-specific mass media.

Estimation of existing radiation risks and their prediction are used for making managerial decisions in relation to radiation exposure and radiation protection of workers. The estimates should be taken into account when planning accident rescue operations and setting up emergency teams.

References

1. V.K. Ivanov, A.F. Tsyb, A.P. Panfilov, A.M. Agapov, O.V. Kaidalov, A.M. Korelo, M.A. Maksioutov, S.Y. Chekin, P.V. Kashcheyeva, Estimation of individualized radiation risk from chronic occupational exposure in Russia, Health Phys. **97**, 2 107-114 (2009).
2. S.G. Mikheenko, et al., ARMIR: the system for estimation of radiological risk from occupational exposure. Proceedings of the International Conference on Occupational Radiation Protection (Vienna, IAEA, 2014).
3. V.K. Ivanov, A.M. Korelo, K.A. Tumanov, S.Yu. Chekin, S.A. Adamchik, S.G. Mikheenko, A.P. Panfilov, V.Yu. Usoltsev, Monitoring of occupational radiation risks of nuclear industry workers (ARMIR system), Radiation and Risk **25**, 1 16-24 (2016).
4. International Commission on Radiological Protection, The 2007 Recommendations of the International Commission on Radiological Protection, Publication 103, Ann. ICRP **37**, 2-4 (2007).

https://doi.org/10.1051/bioconf/20191403021

Comparative analysis of the uranium content in the urine of personnel of group A and persons not exposed

Moldir Aumalikova[1], *Danara* Ibrayeva[1], *Polat* Kazymbet[1], *Meirat* Bakhtin[1], *Yerlan* Kashkinbaev[1], *Maksat* Sharipov[1]

[1]Institute of radiobiology and radiation protection JSC "Astana medical university", Kazakhstan

The hydrometallurgical plant of the Stepnogorsk Mining and Chemical Combine was established in the fifties of the last century. The main activity of the Hydrometallurgical Plant is the extraction and processing of uranium ore.

Uranium content in urine in people whose profession is not related to sources of ionizing radiation from different sources is from 0.04 to 0.4 micrograms per liter [1]. Referring to the US normative guidance 8.22 "Bioanalyses in uranium mines" as a conditional normative value for workers in the uranium industry will be designated 15 μg/l [2].

In this paper, the results of the study of the uranium content in the average daily urine of the staff of the mining and smelting plant in Stepnogorsk are compared with those who do not contact with ionizing radiation. The uranium content in urine was determined by the inductively coupled plasma mass spectrometry model "Agilent 7800 ICP-MS". The results of mass spectrometric analysis showed that the minimum uranium content in urine was 4.5 μg/l, and the maximum value was 28 μg/l, and for workers not working with ionizing radiation the minimum value was 0.05 μg/l, the maximum value was 0,08 μg/l.

We calculated the expected effective dose of internal irradiation from the urine sample analysis for uranium content according to the formula below.

$$D=\frac{T*Aurina}{\sum_{i=1}^{T} m(T+t-i)} * d \quad (1)$$

where, A_{urina} is the activity of uranium isotopes in the daily urine sample, which is calculated by the formula $A = \frac{0.693*m}{T_{1/2}} * \frac{N_A}{\mu}$;

m (T + t-i) - function of excretion of uranium in urine for chronic intake during time T;

t-time elapsed after the end of the receipt until the time of collection of the daily urine sample, provided that the monitoring was carried out 2 times a year;

d - the dose equivalent or effective dose rate for the given pathway.

When calculating the expected effective dose of internal irradiation from the inhalation pathway of uranium intake, the particle sizes (average aerodynamic diameter of activity) equal to 5 μm were taken into account. Due to the lack of information on the types of incoming uranium compounds, the calculation was made for all types of chemical forms: fast (F), medium (M) and slow (S), but the results of the medium soluble uranium type were most suitable for close values [3,4].

The calculation shows that the expected effective dose of personnel depending on the chemical forms and the average uranium content in urine can vary with a minimum value of 4.5 μg/l ~ 10 mSv/year, with a maximum value of 28 μg/l ~ 67 mSv/year.

For workers who do not have contact with ionizing radiation, the expected effective dose at a minimum value of 0.05 μg / L ~ 0.012 mSv / year, with a maximum value of 0.08 μg / L ~ 0.021 mSv / year.

High levels of uranium in the urine of staff, which in comparison with the "nominal rate" exceed by 1.8 times, that when calculating the expected effective dose exceeds the values of the maximum dose of 20 mSv per year by 3 times, indicates the need for further monitoring studies.

References

1. W.S.Snyder, M.J.Cook, E.S.Nasset, L.R.Karhausen, G.P.Howells, I.H.Tripton Report of the Task Group on Reference Man. A report prepared by a task group of Committee 2 of the International Commission on Radiological Protection. Oxford, Pergamon Press, 1975.

2. Regulatory guide 8.22. Bioassay at uranium mills. U.S. nuclear regulatory commission office of nuclear regulatory research, may 2014, revision 2

3. International Commission on Radiological Protection, Individual Monitoring for internal exposure of workers (Revision of ICRP publication No.54) Publication No.78, Ann.ICRP 27 3-4, Elsevier Science Ltd. – Oxford , 1997

4. Assessment of doses to the public from ingested radionuclides. Safety reports series No.14. International Atomic Energy Agency. Vienna: International Atomic Energy Agency,1999.

Tritium and cancer risk: incorporating informative priors from experimental research into epidemiology

David B. Richardson[1]

[1]School of Public Healthn University of North Carolina at Chapel Hill, Chapel Hill, NC, USA

The health effects of exposure to tritium, a radioactive isotope of hydrogen, are of interest for a number of reasons: exposure is widespread, and includes planned and unplanned releases from nuclear sites; tritium is very mobile; and there remains uncertainty about its biological effects. There have been a number of authoritative reviews of tritium and the experimental evidence regarding its biological effects. These have tended to reach similar conclusions, suggesting relative biological effectiveness values in the range of 1.1–5.0 when compared with chronically-delivered gamma rays; a number of reports have concluded that that tritium beta particles should be expected to be at least as biologically effective as gamma rays. Evidence of tritium effects on human health derived through epidemiological studies, while arguably the most direct way to address questions about tritium health effects, unfortunately, is very limited. This has led some authors to contend that this limited epidemiological evidence means it is necessary to derive estimates of tritium risk indirectly from experimental studies. An alternative viewpoint, however, is to acknowledge the limits of the available evidence (human and experimental) and therefore integrate these lines of evidence, proposing that neither on its own is sufficient to address the question, but combined we may usefully address the question. A major limitation of the epidemiological data is that sparse data make it difficult to obtain reliable estimates of association. In settings of sparse data, Bayesian inference with informative priors can be used to improve estimation of associations. What is that prior information, and how should we incorporate it? As reviewed in a recent US NCRP report, a variety of lines of evidence) pertain to the biological effectiveness of tritium in experimental settings. These tend to suggest that low-energy beta emitters will have greater biological effectiveness per unit absorbed dose than standard reference radiations under most circumstances. We have proposed to incorporate such evidence as a prior, implemented with what is known as an ordered constrained prior. Under this prior, we posit that the effect of tritium per unit absorbed dose is at least as effective as external photons (primarily gamma radiation) per unit absorbed dose. We have illustrated this in a study of radiation dose and leukemia mortality among 18,883 workers at the Savannah River Site employed from 1951-1999 and followed through 2002. The approach yields a posterior estimate of the excess relative rate per 100 mGy for external photon exposure and an estimate of the excess relative rate per 100 mGy for tritium intakes. The ratio of these estimates suggests a relative effectiveness of approximately 4, a value that falls within the range of estimates obtained from experimental studies (e.g., 1.1 – 5), and suggests a potentially useful direction for further integration of experimental and epidemiological lines of evidence regarding tritium health effects.

BIO Web of Conferences **14**, 04002 (2019) https://doi.org/10.1051/bioconf/20191404002
HEIR 2018

Cancer risk from exposure to alpha emitters: a view from the ICRP

Margot Tirmarche[1], *Dominique* Laurier[2], *A. Iulian* Apostoaei[3], *Eric* Blanchardon[2], *Elizabeth* Ellis[4], *Ethel* Gilbert[5], *John D.* Harrison[6], *Jean-François* Lecomte[2], *James W.* Marsh[6], *François* Paquet[2], *Mikhail* Sokolnikov[7], *Richard* Wakeford[8], and *Sergei* Zhivin[2,9]

[1] Autorité de Sûreté Nucléaire (ASN), Chatillon, France
[2] Institut de Radioprotection et de Sûreté Nucléaire (IRSN), Fontenay aux Roses, France
[3] Oak Ridge Center for Risk Analysis, Inc., Oak Ridge, TN, USA
[4] Oak Ridge Associated Universities, Oak Ridge, TN, USA
[5] National Cancer Institute (NCI), Rockville, MD, USA
[6] Public Health England, Centre for Radiation, Chemical and Environmental Hazards, Chilton, UK
[7] Southern Urals Biophysics Institute (SUBI), Ozyorsk, Russia
[8] Centre for Occupational and Environmental Health, University of Manchester, Manchester, UK
[9] Institut national de la santé et de la recherche médicale (Inserm), Villejuif, France

A significant number of new results have been published since 2000 from epidemiological studies addressing the risk of cancer associated with internal exposure to radionuclides emitting alpha radiation. The International Commission on Radiological Protection (ICRP) mandated a task group (TG64) to review these results and evaluate their their implications for the protection. TG64 is composed of members from ICRP Committees 1 and 2, as well as external experts with epidemiologic and dosimetric expertise.

The alpha-particle emitting radionuclides considered were radon-222 and its decay products, plutonium-239 and isotopes of uranium.

The presentation will summarize the current state of knowledge regarding cancer risk associated with radon, plutonium and uranium. The main conclusions reached in relation to radon have already been published [1], and will be put in context with the recently published radon dose coefficients [2]. Results obtained on cancer risk associated with plutonium or uranium exposures will be summarized, and current work of TG64 on lifetime risk estimates presented in details. In addition, the lifetime lung cancer risks from inhalation of plutonium are compared with the corresponding risks from low-linear energy transfer (low-LET) radiation, which provides some indication of possible values of the relative biological effectiveness (RBE) of alpha particles for the induction of lung cancer.

References

1. M. Tirmarche, J.D. Harrison, D. Laurier, F. Paquet, E. Blanchardon, J.W. Marsh. Lung Cancer Risk from ICRP Publication 115: Radon and Progeny and Statement on Radon. Ann. ICRP. **40**, 1 (2010).

2. F. Paquet, M.R. Bailey, R.W. Leggett, J. Lipsztein, J.W. Marsh, T.P. Fell, T. Smith, D. Nosske, K.F. Eckerman, V. Berkovski, E. Blanchardon, D. Gregoratto, J.D. Harrison;

Authors on Behalf of ICRP. ICRP Publication 137: Occupational Intakes of Radionuclides: Part 3. Ann. ICRP. **46**, 1 (2018).

BIO Web of Conferences **14**, 04003 (2019) https://doi.org/10.1051/bioconf/20191404003
HEIR 2018

Pooled Uranium Miners Analysis (PUMA): The Setting Up of an International Occupational Cohort

Estelle Rage[1*], *Minh* Do[2], *Paul A* Demers[2], *Nora* Fenske[3], *Michaela* Kreuzer[3], *Dominique* Laurier[1], *Jon* Samet[4], *Mary* Schubauer-Berigan[5], *Christina* Sobotzki[3], *Ladislav* Tomasek[6], *Lydia* Zablotska[7], and *David* Richardson[8]

[1]Institute for Radiation Protection and Nuclear Safety, Fontenay-aux-Roses, France
[2]Occupational Cancer Research Centre, Toronto, Canada
[3]Federal Office for Radiation Protection, Neuherberg, Germany
[4]University of Southern California, Los Angeles, CA, USA
[5]National Institute for Occupational Safety and Health, Cincinnati, OH, USA
[6]Radiation Protection Institute, Prague, Czech Republic
[7]University of California, San Francisco, San Francisco, CA, USA
[8] University of North Carolina, Chapel Hill, NC, USA

Introduction: Epidemiological studies of underground miners have highlighted the increased risk of lung cancer death associated with cumulative radon exposure. These studies contributed to the setting up of radioprotection measures in order to limit the exposure to radon and its short-lived progeny. To date, questions remain regarding the risk of lung cancer associated with chronic exposure to low levels of radon present in occupational and environmental settings. Furthermore the question is raised for the risks of cancers other than lung and for the risks of non-cancer diseases, such as circulatory system diseases. Thereby, the setting up of a large combined cohort gathering some of the most informative cohorts of uranium miners will allow us to conduct epidemiological analyses with an increased statistical power.

Methods: The Pooled Uranium Miners Analysis (PUMA) includes seven cohorts of uranium miners: three European cohorts (Czech, French and German cohort), two Canadian cohorts (Eldorado and Ontario cohort) and two American cohorts (Colorado Plateau and New Mexico cohort). Main inclusion criteria were that all of these miners must have worked in the uranium mining industry and had to be identified in a nonselective way. All members of the cohorts have an individual assessment of their annual exposures to radon. Health information such as vital status and causes of death must be complete.

Results and perspectives: To date, the PUMA study includes more than 120,000 uranium miners and contributes to more than 4.3 million of person-years at risk. Most of the uranium miners were men. Only the German and Eldorado cohort includes women, corresponding to 4% of the PUMA cohort. According to the cohorts, the mean duration of the follow-up stretches from 30 to 39 years and the mean cumulative exposure to radon

[*] Corresponding author: estelle.rage@irsn.fr

varies from 31 to 579 Working Level Month (WLM). The total number of deaths reaches 49,000 cases and more than 7,400 of lung cancer deaths were counted. The PUMA study has been set up to strengthen the knowledge on our understanding of radon and radon progeny-related diseases. On the basis of this large cohort, the coming perspectives are to perform a mortality analysis and to assess the exposure-risk relationships.

Risk of chronic obstructive pulmonary disease in a cohort of workers occupationally exposed to ionizing radiation over prolonged periods

Tamara Azizova[1], *Galina* Zhuntova[1], *Evgenia* Grigorieva[1], *Maria* Moseeva[1], *Maria* Bannikova[1], and *Evgeniy* Bragin[1]

[1]Southern Urals Biophysics Institute, Ozyorsk, Russia

Introduction

Mayak Production Association is a Russian large-scale nuclear enterprise producing weapon-grade plutonium which was set in operation in 1948. Equipment defects and irregularity of technology process, low level of production mechanization and automatization in early years of the enterprise operation caused high levels of alpha-radioactivity in occupational air [1]. As Mayak PA workers did not use individual protection means for organs of respiratory system, alpha-particles of plutonium deposited in workers' organisms via inhalation and were accumulated in main organs of plutonium deposition (lung, liver, bones) [2]. The first plutonium-induced pulmonary fibrosis (an early tissue reaction) was diagnosed in a plutonium production plant worker in 1956 [3].

The present study was aimed to assess the risk of chronic obstructive pulmonary disease (COPD) in workers of the Mayak PA exposed to internally deposited alpha-particles.

Material and methods

The study cohort included Mayak PA workers first employed at one of the main facilities (industrial reactors, radiochemical and plutonium production plants) in 1948–1982, in total 22,377 individuals (with 25% of females).

The follow-up of the cohort started with a date when a worker was hired and lasted until the earliest of the following events: date of COPD diagnosis, date of death, 31.12.2008, date of the last available medical record for migrants and residents with unknown vital status.

By the end of the follow-up vital status data was available for 95.0% of cohort members of whom 53.5% had deceased and 46.5% were alive. Information on disease incidence through the entire follow-up period was collected for 95.0% of workers, data on smoking status was available for 97.0% of workers with quantitative parameters estimated for 51.0% of ever-smoking workers.

To perform analyses, estimates of absorbed external and internal radiation doses were provided by dosimetry system MWDS–2008 [4].

Relative risk (RR) and excess relative risk per unit dose (ERR/Gy) were calculated based on maximum likelihood using AMFIT module of EPICURE software [5]. Differences were considered as statistically significant with $p<0.05$.

Results

The analysis demonstrated that COPD incidence in the Mayak PA worker cohort was significantly associated with sex, attained age, facility type, smoking status and index, calendar period and hazardous production factors which workers had been exposed to before employment at the Mayak PA (coal dust, chemicals, etc.).

The categorical analysis showed an increase of COPD risk estimates with increasing radiation doses, but significant ERR=1.30 95% CI: 1.04, 1.62) was revealed only in workers exposed to internally deposited alpha-particles at lung absorbed doses exceeding 0.1 Gy.

The table below summarizes results of ERR/Gy analysis for COPD incidence following external and internal radiation exposure.

A significant association of COPD incidence risk with dose from internally deposited alpha-particles was shown for Mayak PA workers first employed at one of the main facilities in 1948–1982.

No significant association of COPD incidence was observed with dose from external gamma-rays. The risk gained significance only after exclusion of adjustments for smoking and dose from internal alpha-radiation exposure.

To improve the accuracy of estimates of radiogenic risks for the study cohort, further studies are required to take into account characteristics of alpha-active aerosols used at the plutonium production and radiochemical plants in addition to occupational radiation doses as well as to investigate a combined effect of occupational radiation exposure and smoking.

Conclusion

The study provides evidence to increased risk of COPD in the cohort of workers occupationally exposed to internal alpha-radiation.

Table **COPD incidence risk in relation to external gamma-ray and internal alpha-particle exposures (ERR/Gy)**

	ERR/Gy (95% CI)	
	Внешнее γ-облучение	Внутреннее α-облучение
Follow-up periods (0 year lag) :	over the whole dose range	
1948–2008	0.03 (-0.05; 0.14)	0.79 (0.17; 1.78)
1948–1960	-0.06 (-0.15; 0.08)	0.15 (na; 1.45)
1961–2008	0.10 (-0.02; 0.26)	1.19 (0.32; 2.53)
	1961–2008 follow-up period	
Analyses with imposed lags		
5 years	0,09 (-0,03; 0,25)	1,14 (0,22; 2,61)
10 years	0,08 (-0,04; 0,23)	1,13 (0,15; 2,75)
15 years	0,09 (-0,04; 0,25)	1,28 (0,19; 3,08)
20 years	0,11 (-0,03; 0,29)	1,94 (0,39; 4,50)
Analysis without adjustments for risk factors		
Smoking	0,16 (0,04; 0,32)	0,79 (0,19; 1,69)
Alpha-radiation	0,20 (0,08; 0,34)	0,40 (0,04; 0,93)
Alpha-radiation and smoking	0,24 (0,12; 0,40)	0,39 (0,04; 0,88)
Alpha-radiation, smoking and facility type	0,12 (0,03; 0,22)	0,45 (0,10; 0,94)
Analysis restricted with data on workers of		
Facility type:		
Reactors	0,12 (-0,11; 0,50)	–
Radiochemical plant	-0,09 (-0,26; 0,18)	0,63 (na; 2,46)
Plutonium production plant	0,19 (0,004; 0,48)	1,65 (0,38; 3,97)
	P_1 =0,09	P_1 = 0,36
Sex:		
Males	0,10 (-0,03; 0,27)	1,45 (0,32; 3,20)
Females	0,13 (na; 0,67)	0,74 (na; 3,10)
	$P_2 > 0{,}50$	$P_2 > 0{,}50$
Attained age (years):		
< 40	0,32 (na; 1,24)	0,37 (na; na)
40 – 49	-0,03 (na; 0,27)	0,88 (na; 5,91)
50 – 59	-0,02 (-0,17; 0,23)	1,47 (na; 4,15)
60 +	0,32 (0,03; 0,82)	0,82 (na; 3,85)
	P_3 =0,12	P_3 = 0,36
	P_4 =0,24	P_4 = 0,37
Smoking status:		
Never-smokers	0,10 (na; 0,57)	0,87 (na; 2,96)
Former smokers	0,40 (0,00; 1,28)	2,09 (na; 9,59)
Smokers	0,10 (-0,05; 0,33)	1,30 (na; 3,70)
	$P_5 > 0{,}50$	$P_5 > 0{,}50$
Smoking index (pack-years):		
0 (did not smoke)	0,10 (na; 0,57)	0,87 (na; 2,96)
<10	-0,15 (na; na)	0,94 (na; na)
≥ 10	0,15 (-0,01; 0,37)	1,56 (0,25; 3,64)
	P_6 = 0,37	$P_6 > 0{,}50$
Pre-employment hazardous factors:		
Yes	0,15 (-0,02; 0,40)	0,52 (na; 2,09)
Potentially	0,02 (-0,17; 0,37)	-0,05 (na; na)
No	0,12 (-0,12; 0,53)	4,08 (1,44; 8,54)
	$P_7 > 0{,}50$	P_7 = 0,01

na –95% CI were not defined

P_1 – test for heterogeneity between workers of radiochemical and plutonium production plants;

P_2 – test for heterogeneity between male and female workers;

P_3 – test for heterogeneity among various age groups of workers;

P_4 – test for log-linear trend for ERR/Gy with attained age;

P_5 – test for heterogeneity among groups of workers with various smoking statuses;

P_6 – test for heterogeneity among groups of workers with various smoking indexes (< 10 и ≥ 10 pack-years);

P_7 – test for heterogeneity between groups of workers exposed and unexposed to hazardous industrial factors prior to Mayak PA employment.

References

1. A. Kruglov, The history of the Soviet atomic industry Taylor and Francis, 2002.
2. L. Ilyin, Plutonium. Radiation Safetyt Mascow, Izdat, 2005.
3. A. Mishachev, Revisiting an issue of occupational pneumosclerosis in workers of plutoniu-239 processing facilities. Bulletin of Radiation Medicine. 1, 10-17 (1963) [in Russian]
4. V. Khokhryakov, V. Khokhryakov, K. Suslova, V. Vostrotin, V. Vvedensky, A. Sokolova, et al Mayak Worker Dosimetry System 2008 (MWDS-2008): Assessment of internal alpha-dose from measurement results of plutonium activity in urine. Health Phys 104: 366–378 (2013).
5. D. Preston, J. Lubin, D. Pierce, M. McConney, *Epicure Users Guide*. Hirosoft, Seattle, WA (1993).

Mortality from cancer (other than lung) and radon exposure in German uranium miners, 1946–2013

Nora Fenske[1,*], *Veronika* Deffner[1], *Maria* Schnelzer[1], *Christina* Sobotzki[1], and *Michaela* Kreuzer[1]

[1]Federal Office for Radiation Protection (BfS), Division Effects and Risks of Ionising and Non-Ionising Radiation, Neuherberg, Germany

1 Background

Radon is an established risk factor for lung cancer. Less clear is whether radon can cause cancers other than lung cancer. Due to the several to many times lower organ doses for organs other than the lung, the radon-related risks - if any - are expected to be much lower than the effect of radon on lung cancer. Therefore, large study populations with a wide range of exposures and long observation periods are required to investigate and to provide evidence for such associations. Based on the current follow-up period (1946–2013) of the German uranium miner cohort study (n=58,974 men) the association between mortality from cancers other than lung cancer and cumulative exposure to radon and its progeny (mean=280, median=33, max=3224 WLM) was investigated.

2 Methods

Annual exposures to radon and its progeny in working level months (WLM) and to external gamma radiation in mSv were retrospectively assessed using a comprehensive job-exposure matrix. Considered endpoints were the group of non-lung cancers as well as all individual cancer sites with more than 35 deaths. Standardized mortality ratios were calculated to compare cancer mortality rates observed in the cohort with those of the general male population in Eastern Germany. Statistical modelling of the mortality rates in the cohort was performed by internal Poisson regression for grouped data with baseline stratification by age and calendar year. Excess relative risks (ERR) per 100 WLM and 95% confidence intervals (CI) were estimated for cumulative 5-year lagged exposure to radon. Numerous additional and sensitivity analyses were conducted, such as additive adjustment for cumulative exposure to external gamma radiation, modelling of non-linear associations, modelling of effect modifications by time-dependent factors, or additional stratification for workplace and duration of employment.

* Corresponding author: nfenske@bfs.de

3 Results

A statistically significant association was found for the group of non-lung cancers (n=5,122; ERR/100 WLM=0.014; 95% CI: 0.006, 0.021) which did not change after adjustment for external gamma radiation. Additional analyses indicated an effect modification depending on age at exposure. With respect to individual cancer sites, the majority of risk estimates were positive (14 out of 18), and two of them reached borderline significance (colon and liver cancer), even after adjustment for external gamma radiation. The largest ERR coefficients were estimated for pharynx, liver and non-Hodgkin lymphoma.

Overall, the results are in line with those based on previous follow-up periods [1-5], but the increased statistical power leads to smaller confidence intervals for estimated ERR coefficients (in comparable statistical analyses [1, 2]).

4 Conclusion

The present analysis provides some evidence of an increased risk for cancers other than lung cancer associated with radon, but confounding cannot be ruled out. Limiting factors of the analysis are the lack of data on cancer incidence, lack of information on potential confounders (such as alcohol consumption and smoking), and the uncertainty of exposure in early years. Several projects are ongoing to address these drawbacks: an international pooling of uranium miner studies (PUMA), a linkage with the German cancer registry, and uncertainty analyses in the German uranium miner cohort.

References

1. M. Kreuzer, L. Walsh, M. Schnelzer, A. Tschense, B. Grosche, Br. J. Cancer **99** (2008)
2. M. Kreuzer, B. Grosche, M. Schnelzer, A. Tschense, F. Dufey, L. Walsh, Radiat. Environ. Biophys. **49** (2010)
3. M. Kreuzer, K. Straif, J.W. Marsh, F. Dufey, B. Grosche, D. Nosske, M. Sogl, Occup. Environ. Med. **69** (2012)
4. F. Dufey, L. Walsh, M. Sogl, A. Tschense, M. Schnelzer, M. Kreuzer, J. Radiol. Prot. **33** (2013)
5. M. Kreuzer, F. Dufey, J.W. Marsh, D. Nowak, M. Schnelzer, L. Walsh, Int. J. Rad. Biol. (2014)

Lung, liver and bone cancer lifetime risk after intake of ^{239}Pu

Mikhail Sokolnikov[1], *Aleksandr* Yurkin[1], *Vadim* Vostrotin[1], *Aleksandr* Ephimov[1], *Evgeny* Vasilenko[1], and *Sergey* Romanov[1]

[1]Southern Ural Biophysics Institute, Ozyorsk, Russian Federation

Workers involved in processing of ^{239}Pu are subject to two major intake pathways of the nuclide: inhalation and wounds. While the inhalation intake limit is based in understanding of Pu biokinetics and lung clearance, the intake limits for intake via wounds do not exist.

Three organs in the human body are subject to greatest exposures from incorporated Pu: lung, liver and bone after inhalation intake and liver bone after intake through wounded skin.

Earlier we have calculated excess risk of lung, liver and bone cancer related to inhalation intake of ^{239}Pu in Mayak Worker Cohort. In our current report we use these models to calculate lifetime risk of lung, liver and bone cancer after ^{239}Pu intake via inhalation and wounded skin.

In order to do this, we calculated equivalent doses resulting from ^{239}Pu incorporation in lung, liver and bone according to several scenarios including (among others) chronic inhalation intake at the level of annual limit of intake (ALI). We used Lifetime Excess Risk (LER) and Risk of Exposure Induced Death (REID) and their annual increments as characteristics of excess lifetime detriment caused by incorporated ^{239}Pu.

For the scenario of chronic intake at the ALI level we demonstrate that at age 70 the annual lifetime risk increment of lung cancer is by a factor of 4 higher than that assumed by regulatory bodies as acceptable (0.001 per year).

We discuss the reasons for that and possible ways of avoiding extra carcinogenic risk in workers who contact to ^{239}Pu or other transuranium nuclides with long biological half-life.

Association analysis between mortality and ionizing radiation exposure in French uranium workers

Ségolène Bouet[1], *Estelle Davesne*[2], *Eric Samson*[1], *Iris Jovanovic*[3], *Eric Blanchardon*[2], *Cécile Challeton - de Vathaire*[2], *David B Richardson*[4], *Klervi Leuraud*[1], *Dominique Laurier*[5], *and Olivier Laurent*[1*]

1 Institut de Radioprotection et de Sûreté Nucléaire (IRSN), Laboratoire d'Epidémiologie des rayonnements ionisants (PSE-SANTE/SESANE/LEPID), Fontenay-aux-Roses, France
2 Institut de Radioprotection et de Sûreté Nucléaire (IRSN), Laboratoire d'Evaluation de la Dose Interne (PSE-SANTE/SDOS/LEDI), Fontenay-aux-Roses, France
3 Amplexor, Montigny-le-Bretonneux, France
4 University of North Carolina, Chapel Hill, USA
5 Institut de Radioprotection et de Sûreté Nucléaire (IRSN), PSE-SANTE/SESANE, Fontenay-aux-Roses, France

To investigate associations between mortality and chronic internal and external exposure to ionizing radiation in a cohort of nuclear workers with potential for internal exposure to uranium.

Workers employed for at least six months in five plants involved in the French nuclear fuel cycle were included and followed up for mortality between 1968 and 2013. Cause-specific standardized mortality ratios were calculated for all workers. Internal and external absorbed doses were reconstructed for each worker with computerized information. Analyses of associations between radiation dose and cause-specific mortality were conducted using Poisson regression.

The cohort includes 4541 workers. At the end of the follow-up, 838 workers were deceased and 28 lost to follow-up. A healthy worker effect was observed. The risk of prostate cancer mortality was significantly higher for workers exposed to cumulative external dose higher than 50 mGy as compared to non-exposed, but this association became non-significant after adjustment for smoking. Associations between mortality and internal dose showed no consistent pattern. Except for prostate cancer mortality, adjustments for smoking and other risk factors rarely documented in cohorts of radiation workers had no marked influence on findings.

It is the first time in France that such a study was conducted in a cohort of uranium workers with a complete reconstruction of internal dose. Results should be considered as preliminary and interpreted with caution because of the limited size of the cohort and significant sources of uncertainty, notably on dose assessment. Future steps of this study will allow overcoming these limitations.

---

*Corresponding author: olivier.laurent@irsn.fr

Cancer incidence and mortality among uranium miners in the Příbram region of the Czech Republic

Kaitlin Kelly-Reif [1*], *Dale* Sandler[2], *David* Shore[2], *Mary* Schubauer-Berigan [3], *Melissa* Troester [1], *Leena* Nylander-French, [1], and *David* Richardson, [1]

[1]University of North Carolina, School of Public Health, 135 Dauer Drive, NC 27599, USA
[2] National Institute of Environmental Health Sciences, 111 TW Alexander Dr RTP, NC 27709, USA
[3] National Institute of Occupational Safety and Health, 5555 Ridge Road Cincinnati, OH 45213, USA

1 Introduction

It is well established that underground miners exposed to radon progeny experience higher rates of lung cancer [1]. Several cohort studies of uranium miners have confirmed this, although magnitudes of associations vary somewhat between studies. Studies also suggest that radon may be associated with cancer types other than lung [1]. But, many of these findings are based on external comparison populations and may not evaluate individual radon exposure. More internal studies of dose-response relationships between radon and cancer subtypes are necessary. We investigated several subtypes of cancer incidence and mortality among underground uranium miners exposed to low and moderate levels of radon progeny.

2 Methods

2.1 Cohort and case cohort studies

A total of 16,434 male employees who worked at least 12 months underground between 1946 and 1991, and were alive and residing in Czechoslovakia on January 1, 1977 were included in the follow-up cohort [2]. Miners contributed person-time from the start of follow-up (1/1/1977) until the earliest of the date of death among deceased miners, date of migration out of the Czech Republic, or end of the study period (12/31/1992). Radon exposure in cumulative working level months (WLM) was estimated using a job exposure matrix based on company work records and area radon monitors.

A case cohort study was developed to collect additional covariate information including smoking and more precise radon doses [2]. A subcohort of 2,393 subjects was selected from the cohort by stratified random sampling on age at start of follow-up. Follow up for cancer incidence was extended through 1996.

2.2 Statistical analysis

* Corresponding author: kaitlink@live.unc.edu

Standardized Mortality Ratios (SMRs), Standardized Incidence Ratios (SIRs), and associated 95% confidence intervals (95%CI) were estimated by fitting a lognormal Poisson model to calculate expected cases. Expected rates were based on the national population of Czech males specific to 5-year age groups and annual calendar periods.

For internal analyses in the cohort and case cohort study, the relationship between cumulative radon exposure and cancer deaths of interest was modeled using the general model form in equation 1.

$$rate=exp(a_0+\sum_{j=2}^{p} a_j x_j)(1+a_1 d) \quad rate=exp(a_0+\sum_{j=2}^{p} a_j x_j)(1+a_1 d)$$

$$rate=exp(\beta_0+\sum_{i=1}^{k-1}\beta_i d_i+\sum_{j=k}^{p-1}\beta_j x_j) \quad rate = \exp(\beta_0 + \sum_{i=1}^{k-1} \beta_i d_i + \sum_{j=k}^{p-1} \beta_j x_j) \tag{1}$$

α_1 is the excess relative rate (ERR) per unit of lagged cumulative WLM, *d*. α_j are parameters for effects of covariates x_j. Rate ratios by WLM category were also examined.

3 Results

There was a 52% increase in deaths from all malignant causes of death compared to expected rates (SMR = 1.52 95%CI: 1.44-1.60). Lung cancer mortality and incidence were both substantially elevated (SMR = 2.12, 95%CI: 1.96-2.28; SIR = 2.31, 95%CI: 2.15-2.48). Stomach cancer (SIR = 1.37, 95%CI: 1.11-1.63), rectal cancer (SIR = 1.41, 95%CI: 1.16-1.66), liver cancer (SIR = 1.70, 95%CI: 1.16-2.25), and extrathoracic cancer incidences (SIR = 1.41, 95%CI: 1.15-1.77) were also in excess.

Lung cancer mortality increased with higher cumulative radon exposure (ERR/100WLM = 0.22, 95%CI: 0.10-0.37). Windows of exposure varied substantially, with the highest ERR in the 15-30-year window (ERR/100WLM = 0.44, 95%CI: 0.21-0.67). Positive but imprecise associations between cumulative radon exposure and extrathoracic airway cancers were observed (ERR/100 WLM = 0.12, 95%CI: -0.25-0.48), as well as chronic lymphocytic leukemia (CLL) (ERR/100 WLM = 0.24, 95%CI: -0.80-1.27).

In the case cohort study, the estimated linear excess relative rate of lung cancer incidence per 100 WLM was modified by smoking status. The estimated ERR/100 WLM among non-smokers was 0.12 (95%CI: -0.09, 0.33) and among smokers 1.34 (95%CI: 0.88, 1.80).

4 Conclusions

The Příbram miner study is a large, well enumerated cohort with available data on incidence and mortality, as well as smoking for some workers. Our results support the positive association between radon progeny and lung cancer. The results also suggest that more research is needed on prolonged exposure to low levels of radon and death from cancers other than lung, especially extrathoracic cancer and CLL.

This work was supported by the National Institute for Occupational Safety and Health, Centers for Disease Control and Prevention (T42-OH008673) and was funded in part by the Intramural Research Program of the NIH, National Institute of Environmental Health Sciences (Z01-ES049028).

References

1. National Research Council. *Health Effects of Exposure to Radon*. Washington, D.C.:

National Academies Press, 1999.
2. Kulich M, Řeřicha V, Řeřicha R *et al.* Incidence of non-lung solid cancers in Czech uranium miners: a case-cohort study. *Environ Res* 2011;**111**:400–5.

Risk of Leukaemia in Czech Uranium Miners

Ladislav Tomasek[1*], *Robert* Řeřicha[2]

[1]National Radiation Protection Institute, Prague, Czech Republic
[2]Center of Epidemiological Studies, Příbram, Czech Republic

Introduction: Epidemiological studies among uranium miners have evidenced lung cancer risk in relation to radon exposure. Although leukemia is especially sensitive to induction by radiation, the assessment of leukemia risk in uranium miners is more complex as its incidence is substantially lower than that of lung cancer. One of first results of leukaemia risk was reported by Řeřicha in 2006 and was based on case cohort approach with 84 leukaemia cases analysed in relation to cumulated exposure of radon decay product. However, the assessment of the risk is more complex because the radiation risks in mines involve exposure to radon, external gamma radiation, and exposure to airborne long lived radionuclides arising from uranium ore dust.

Results: Recent follow-up of a Czech study of 10 000 uranium miners have resulted in 54 leukaemias. The risk is evaluated in relation to cumulated equivalent doses from radon, long lived radionuclides (LLR), and external gamma. The mean equivalent doses to the red bone marrow in the entire study are 33 mSv from radon, 46 mSv from external gamma radiation, and 106 mSv from long lived radionuclides. Mean doses among leukaemia cases are nearly twice. The excess relative risk per Sv (ERR/Sv) in this study is 3.8 (90% confidence interval 1.0 – 10.5) is statistically significant (p=0.008). This value is consistent with the leukaemia risk observed in the Life Span Study among Japanese survivors of A bombing (ERR/Sv = 4). These results will be compared to earlier results by Řeřicha using the same approach in relation to exposure to radon, external gamma radiation, and exposure to long lived radionuclides.

The study was supported by the Ministry of the Interior of the Czech Republic, project No. MV-25972-2/OBV-2012-2017.

* Corresponding author: ladislav.tomasek@suro.cz

Analysis of histological subtypes of incident lung cancer among Eldorado uranium workers

*1*Lydia B.* Zablotska

[1]School of Medicine, University of California, San Francisco, San Francisco, CA,94143, USA

Exposure to radon decay products (RDP) is one of the best-studied carcinogenic phenomena in radiation epidemiology [1]. Epidemiological studies, primarily of underground miners, show increases in lung cancer risk from exposure to RDP [1, 2]. Only a few studies have been published on the radiation risks of specific histological subtypes of lung cancer [3-6]. Radon exposure in uranium miners seems to be especially related to small cell and squamous cell lung cancer [4]. The latter one follows the pattern of distribution of radon decay products in the bronchial epithelium, from which it originates. Tobacco smoking has been shown to be more strongly associated with squamous cell and small cell carcinoma than with adenocarcinoma [4, 7, 8]. The manner of interaction between smoking and RDP exposures in causing lung cancer is a matter of debate [3, 9, 10].

This study assessed the relationship between RDP exposure and lung cancer incidence in a cohort of 16,770 Eldorado uranium workers from Canada first employed in 1932-1980 and followed up for cancer incidence through the Canadian Incidence DataBase (CIDB) during 1969-1999. The current analysis was based on substantially revised identifying information and RDP dosimetry for workers from the Beaverlodge and Port Radium uranium mines, and from a radium and uranium refinery and processing facility in Port Hope. In brief, the annual mean RDP exposure in working level months (WLM) was calculated by summing over the WL measurements available for each type of workplace, the proportion of workers in each occupation and the proportion of time spent in each type of workplace by workers in each occupation. Workplace RDP concentration measurements were supplemented by data on seasonal mine ventilation rates, building air volumes and air exchange rates. The current study also has information on individual γ-ray doses for all cohort subjects. Personal γ-ray doses were calculated from the average dose-rates and time on the job and expressed in millisieverts (mSv) for workers who did not wear a personal dosimeter.

Each individual contributed person-years at risk from the later of the date of hire or the start date of follow-up, January 1st, 1969. 'Exit date' was defined as December 31st, 1999, the date of cancer diagnosis or death, or the last date known alive (defined as date of last employment or contact), whichever occurred earlier.

The initial set of analyses was based on external comparisons of the cohort with the general Canadian population. Observed and expected values were used to estimate standardized incidence ratios (SIR) by means of indirect standardization. Expected values were derived from Canadian population cancer incidence (1969–1999) rates, adjusted for

*Corresponding author: Lydia.Zablotska@ucsf.edu

sex, age and calendar year at risk. In addition, we separately estimated SIRs for major histological subtypes of lung cancer (squamous cell, small cell, adenocarcinoma, other) for 1985-1999. Poisson regression was used to estimate radiation risks of lung cancer (together and separately for major histological subtypes) from RDP exposures and gamma-ray doses and to investigate possible modifying effects of time since exposure, exposure rate and age at risk. All analyses were adjusted for age at risk (15–19, 20–24… 85–100 years), calendar year at risk (1969-1974, 1975-1979… 1995–1999), employment site (Port Hope, Port Radium, Beaverlodge, "other sites"), total duration of employment (0-6 months, 6 months+), age at first exposure, cumulative exposure, and years since first exposure. All RDP exposures and γ-ray doses were lagged by 5 years to account for the latency period between exposure and lung cancer incidence. Regression parameters, 95% confidence intervals (95% CI) and p-values were estimated using the maximum likelihood method in the AMFIT module of the EPICURE software [11]. Tests of statistical significance were based on the likelihood ratio test and all p-values were two-sided.

Incidence rates of lung cancer were significantly higher compared to the Canadian national calendar time-, sex- and age-specific rates. SIRs increased with increasing RDP exposures in a dose-dependent way. Sub-group analyses showed that significantly elevated SIRs were observed only for workers from the Beaverlodge and Port Radium facilities (uranium miners), whiles SIRs for workers from Port Hope facility (uranium millers) were elevated but not statistically significant and SIRs for workers employed at "other sites" were significantly lower compared to the Canadian general population rates of lung cancer.

Most workers were male (N=15,336, 91.6 % of the cohort), so all further analyses were restricted to "all male workers." During follow-up, 626 lung cancers were diagnosed in the cohort (33% squamous cell, 16% small cell, 18% adenocarcinoma and 33% other subtypes). We observed a monotonic increase in risk for lung cancer incidence, with the highest risks seen in the highest RDP exposure category. The estimated excess relative risk for RDP exposures was positive and statistically significant ($p<0.001$), with no estimated differences in the dose-response between different facilities (p heterogeneity = 0.38). Estimated radiation risks for squamous cell subtype were significantly different from risks for all other subtypes combined (p interaction<0.05). Radiation risks for small cell carcinoma and adenocarcinoma were also significantly increased, but with lower point estimates. We also examined effect modification by time since exposure ($p<0.001$), exposure rate and age at risk (both $p>0.1$). Gamma-ray doses did not significantly improve the fit of the model with RDP exposures.

In summary, the current study supports the use of BEIR VI-type models to predict future risk of lung cancer from past and current RDP exposures. Radiation risks of lung cancer differ by histological subtype. Further studies are needed to better understand the association between affected tissues and radiation risks of lung cancer.

References

1. NRC: Committee on Health Risks of Exposure to Radon. Health effects of exposure to radon. Biological Effects of Exposure to Ionizing Radiation: BEIR VI. (National Research Council (NRC) ed. Washington, DC: National Academies Press; (1999)
2. M. Al-Zoughool, D. Krewski. Health effects of radon: A review of the literature. Int J Radiat Biol. **85**(1):57-69; (2009)
3. H. Egawa, K. Furukawa, D. Preston, S. Funamoto, S. Yonehara, T. Matsuo, S. Tokuoka, A. Suyama, K. Ozasa, K. Kodama, K. Mabuchi. Radiation and smoking effects on lung cancer incidence by histological types among atomic bomb survivors. Radiat Res. **178**(3):191-201; (2012)
4. M. Kreuzer, K. M. Muller, A. Brachner, M. Gerken, B. Grosche, T. Wiethege, H. E. Wichmann. Histopathologic findings of lung carcinoma in German uranium miners. Cancer. **89**(12):2613-21; (2000)
5. D. Taeger, A. Fritsch, T. Wiethege, G. Johnen, A. Eisenmenger, H. Wesch, Y. Ko, S. Stier, K. Michael Muller, T. Bruning, B. Pesch. Role of exposure to radon and silicosis on the cell type of lung carcinoma in German uranium miners. Cancer. **106**(4):881-9; (2006)
6. L. Tomasek, V. Placek. Radon exposure and lung cancer risk: Czech cohort study. Radiat Res. **152**(6 Suppl):S59-63; (1999)
7. S. A. Khuder. Effect of cigarette smoking on major histological types of lung cancer: a meta-analysis. Lung Cancer. **31**(2-3):139-48; (2001)
8. B. Pesch, B. Kendzia, P. Gustavsson, K. H. Jockel, G. Johnen, H. Pohlabeln, A. Olsson, W. Ahrens, I. M. Gross, I. Bruske, et al. Cigarette smoking and lung cancer--relative risk estimates for the major histological types from a pooled analysis of case-control studies. Int J Cancer. **131**(5):1210-9; (2012)
9. K. A. L'Abbe, G. R. Howe, J. D. Burch, A. B. Miller, J. Abbatt, P. Band, W. Choi, J. Du, J. Feather, R. Gallagher, et al. Radon exposure, cigarette smoking, and other mining experience in the beaverlodge uranium miners cohort. Health Phys. **60**(4):489-95; (1991)
10. K. Leuraud, M. Schnelzer, L. Tomasek, N. Hunter, M. Timarche, B. Grosche, M. Kreuzer, D. Laurier. Radon, smoking and lung cancer risk: results of a joint analysis of three European case-control studies among uranium miners. Radiat Res. **176**(3):375-87; (2011)
11. D. L. Preston, J. H. Lubin, D. A. Pierce, M. E. McConney: *EPICURE User's guide*. Seattle, WA: Hirosoft International Corporation; (1993)

Bringing up to date the French database of nuclear workers contaminated with plutonium and/or americium and treated with Ca-DTPA

Stephanie Lamart[1,*], *Lise* Carbone[2], *Laurence* Jossoud[3], *Philippe* Bérard[3], *Xavier* Castagnet[4], *Jean-Vincent* Dauteuil[3], *Gérald* Gagna[5], *Valérie* Juve[6], *Josseline* Lecoix[7], *Yannick* Lecompte[5], *Dominique* Leray[8], *Xavier* Millot[9], *Isabelle* Puech[10], Mariette Ruffin[11], *Michel* Tricotti[12], and *François* Pic[3]

[1]Laboratoire de Radio Toxicologie, CEA, Université Paris-Saclay, 91297 Arpajon, France
[2]Service de Santé au Travail, CEA, 30207 Bagnols-sur-Cèze, France
[3]Cellule Conseiller Médical, CEA, 92265 Fontenay-aux-Roses, France
[4]Service de Santé au Travail, CEA, 33116 Le Barp, France
[5]Service de Protection Radiologique des Armées, 92190 Clamart, France
[6]Service de Santé au Travail, CEA, 13108 Saint-Paul-les-Durance, France
[7]Laboratoire de biologie médicale, CEA, 30207 Bagnols-sur-Cèze, France
[8]Service de Santé au Travail, ORANO, 50440 Beaumont Hague, France
[9]Laboratoire de biologie médicale, CEA, 91191 Gif-sur-Yvette, France
[10]Service de Santé au Travail, CEA, 21120 Is-sur-Tille, France
[11]Laboratoire de biologie médicale, CEA, 13108 Saint-Paul-les-Durance, France
[12]Laboratoire de biologie médicale, CEA, 21120 Is-sur-Tille, France

The diethylenetriaminepentaacetic acid (DTPA) is a chelating agent used to decorporate plutonium (Pu), americium (Am) and curium (Cm) after internal contamination. The DTPA-actinide chelate is mainly excreted in urines. Hence, DTPA decorporation therapy increases and accelerates actinide excretion, which results in a decrease in the actinide systemic retention. In France, Ca-DTPA has been used since 1962 and authorized by the French National Agency for Medicines and Health Products Safety (ANSM) since 2008 [1]. This authorization has been given upon the completion of a first study of French nuclear workers contaminated with Pu and/or Am and treated with injection(s) of DTPA. Data on contamination events, subsequent treatments and bioassays were collected and stored in a database. Main results of the data analysis included:

- descriptive statistics on the 548 collected events of contamination and corresponding treatments protocols, which resulted in a total of 1158 injections,
- an evaluation of the increase of urinary excretion,
- a description of adverse reactions (cutaneous allergy) observed on only one case out of the 469 persons treated
- a common treatment protocol,
- a CEA report and several publications in peer-reviewed journals [1-5].

However, the systematic storage in this database of information on the monitoring and treatment of contaminated workers has been discontinued since the end of the first study in 2004. Still, while DTPA is the treatment of choice after Pu, Am and/or Cm contamination,

* Corresponding author: stephanie.lamart@cea.fr

it is necessary to keep on updating statistics on contamination conditions, treatment protocols, bioassay results and eventual adverse reactions. This information is crucial to optimize treatment protocols and to possibly adapt the protocol on the contamination conditions. It is also a solid foundation for communicating with workers in case a treatment is recommended. Furthermore, the interpretation of excretion data after DTPA treatment remains a challenging step in the estimation of internal dose. In the meantime, biokinetic models from the International Commission for Radiological Protection does not account for DTPA treatment. Therefore, collection of excretion data affected by DTPA treatment can contribute to current efforts for developing a biokinetic model that includes the effect of DTPA.

As additional cases of contamination and DTPA treatments have been observed in France in the past 14 years, it was therefore decided to update this database and re-initiate a study including these new data. A working group composed of occupational physicians, biologists, pharmacists, experts in internal dosimetry and data analysists from the French nuclear industry, has been created early 2018. While the objectives will be similar as the initial study, the approach will be completed. Whereas the first study reported measurement results only based on urine data, the whole set of radiotoxicological and *in vivo* measurement will be used. In particular, faecal measurements will be included as DTPA also results in an increase of faecal excretion, whereas this effect has not yet been thoroughly evaluated. The increase of urinary and faecal excretion after DTPA will be quantified and evaluated depending on the radioisotope and route of entry. A numerical code based on the R-project for statistical computing (www.r-project.org) will be used to carry out this analysis which will facilitate the traceability of the results.

References

1. L. Grappin and P. Bérard, Radioprotection. **43**(3): p. 465-466 (2008).
2. L. Grappin, P. Bérard, P. Beau, L. Carbone, X. Castagnet, C. Courtay, J. Le Goff, F. Ménétrier, M. Néron, and J. Piechowski. Rapport CEA-R-6097 (2006).
3. L. Grappin, P. Bérard, F. Ménétrier, L. Carbone, C. Courtay, X. Castagnet, J.P. Le Goff, M.O. Néron, P. Beau, and J. Piechowski, Radioprotection. **42**(2): p. 163-196 (2007).
4. L. Grappin, P. Berard, F. Menetrier, L. Carbone, C. Courtay, X. Castagnet, J.P. Le Goff, M.O. Neron, and J. Piechowski, Radiat Prot Dosimetry. **127**(1-4): p. 435-9 (2007).
5. L. Grappin, J.-P. Legoff, L. Carbone, C. Courtay, A.-L. Agrinier, M. Aninat, J.-C. Amabile, A. Florin, and F. Andre, Radioprotection. **44**(4): p. 447-461 (2009).

https://doi.org/10.1051/bioconf/20191404012

Descriptive characteristics of the cohort of workers from the Siberian Group of Chemical Enterprises (SGCE)

Andrey Karpov[1,*], *Ravil Takhauov*[1], *Andrey Zerenkov*[1], *Yuriy Dolgopolov*[2], *Konstantin Izmestev*[2], *Aleksey Blinov*[1], *Lydia Zablotska*[3]

[1] Seversk Biophysical Research Center of the Russian Federal Medical and Biological Agency, Seversk, Russia
[2] Siberian Group of Chemical Enterprises of State Corporation Rosatom, Seversk, Russia
[3] School of Medicine, University of California, San Francisco, USA

Siberian Group of Chemical Enterprises (SGCE) is a large complex for production of nuclear fuel which includes multiple processing facilities, including isotope separation and sublimation. The Isotope separation plant and the Sublimation plant were put into commercial operation in 1953 and in 1954, respectively. They produce uranium enriched by isotope U-235 and uranium hexafluoride. Both natural and regenerated uranium are used as raw materials.

Systematic monitoring of uranium alpha-emitting radionuclides in the SGCE employees was initiated in the mid-1950s by specialized biophysical laboratory using the indirect method based on the radiochemical analysis of biological samples. Biophysical examination of the uranium body burden was conducted once per year.

The cohort of SGCE workers potentially exposed to uranium includes all employees who were hired during January 1, 1950-December 31, 2010 and worked for at least 1 year (~ 3,500 workers). Workers from this group may have been exposed to both external radiation (approximately 60 %) and to internal radiation from incorporated plutonium. Table 1 shows descriptive characteristics of SGCE employees potentially exposed to uranium. Follow up was conducted through 2014 through a specialized medical registry for SGCE employees (vital status known for ~ 75 %). Table 2 shows the distribution of uranium group employees by uranium content in urine and cumulative dose of external exposure.

Table 1. Descriptive characteristics of SGCE workers potentially exposed to uranium

* Corresponding author: sbnc@fmbamail.ru

first employed 1950 to 2010.

Variable	Sex	Radio-chemical plant	Plutonium plant	Plutonium plant	Sublimate plant	Support facility	Total
Total number of employees	M/F	154/58	1,664/464	311/133	506/122	89/53	
	Total	212	2,128	444	628	142	**3,554**
Outpatient examination	M/F	35/7	1,349/93	219/108	315/78	50/35	
	Total	42	1,442	327	393	85	**2,289**
Hospital examination	M/F	124/52	859/192	70/18	269/56	53/19	
	Total	176	1,051	88	325	72	**1,712**
Uranium body burden	M/F	25/6	389/97	41/21	94/9	13/5	
	Total	31	486	62	103	18	**700**
Monitored for external exposure	M/F	199/51	975/168	85/12	418/87	56/28	
	Total	250	1,143	97	505	84	**2,079**
Cancer patients: total number (%)	Total	32 (15.1)	289 (13.6)	99 (22.3)	116 (18.5)	28 (19.7)	**564 (15.9)**

Table 2. Distribution of SGCE workers potentially exposed to uranium by categories of uranium content in urine and by cumulative dose of external exposure.

Variable	Uranium content in urine (kBq)				
	< 0.046	0.046-0.74	> 0.74-1.48	> 1.48-3.70	> 3.70
Male	883	1,395	133	47	30
Female	256	448	21	7	12
Total	**1,139**	**1,843**	**154**	**54**	**42**

Index	Cumulative Dose of External Exposure (mSv)						
	0	> 0-50	> 50-150	> 150-200	> 200-300	> 300-500	> 500-1,000
Male	1,068	722	408	113	167	152	87
Female	491	189	89	25	23	11	2
Total	**1,559**	**911**	**497**	**138**	**190**	**163**	**89**

In summary, we presented a first description of the cohort of workers employed at the SGCE who were exposed to a complex combination of external and internal irradiation and potentially could have been exposed to uranium. The electronic database contains detailed information on occupational activities including the information on the measured doses of external exposure and dynamics of their accumulation, as well as the data on biophysical studies for detection of uranium body burden. The cohort has been followed up for mortality and cancer incidence for over 65 years and presents a unique opportunity for conducting research on the medical and biological effects of low-dose chronic radiation exposure. This cohort of SGCE employees is uniquely suited for research on the effects of the long-term low-dose occupational radiation exposures.

Risk assessment of thyroid cancer incidence among population due to residence close to Mayak PA

Irina A. Martinenko[1], *Mikhail E.* Sokolnikov[1], and *Nina A.* Koshurnikova[1]

[1] Southern Urals Biophysics Institute, 456780, Ozyorskoye shosse, 19, Ozyorsk,

Chelyabinskaya oblast, Russia

Preface: It is well known that gas-aerosol emissions from the pipes of nuclear reactors and irradiated fuel disposal plants are the main source of I^{131} pollution of the environment [1]. Due to this fact assessing the risk of health effects, primarily cancer, among the populations residing close to atomic production facilities that used to be a source of radioactive pollution of adjacent areas exceeding the background level is still a topical problem [2].

Topicality of cancer incidence studies among the residents of Ozyorsk, as the nearest settlement to Mayak PA, is supported by the fact that uncontrolled gas-aerosol emissions to the atmosphere from the pipes of production reactors and of radiochemical plant took place in the initial years of operation of the enterprise. It was stated that ^{131}Iwas the main dose-forming radionuclide in release composition [3]; ^{131}I was produced in the course of reprocessing of irradiated uranium lumps. Radioactive iodine was released to the atmosphere with gas-aerosol emissions through air systems of the radiochemical plant that lead to radiation pollution of the adjacent areas including Ozyorsk. In [4] it is demonstrated that the total volume of ^{131}I released to the atmosphere with gas-aerosol emissions through the pipes of the radiochemical plant made 1.0 million Ci, 90% of that volume was released in 1948–1954.

In earlier studies it was demonstrated that increased incidence of thyroid cancer is the only reliable health effect among the population that used to reside in the city in childhood [5].

Purpose: Thyroid cancer incidence rates in the population residing as children in areas contaminated by radioactive gas-aerosol emissions during technological development at Mayak PA were assessed in the current epidemiological study.

Material and methods: The study was carried out in a cohort of 31,836 people residing in childhood in Ozyorsk during 1948–1962. In that period uncontrolled gas-aerosol emissions from chimneys of radiochemical plants took place. There were 60 thyroid cancer cases diagnosed and histologically verified in the study cohort. Standardized incidence rates (SIR) was estimated by the indirect standardization by age and sex with a 95% confidence interval. Age- and sex-related thyroid cancer incidence rates in Russia and among the urban population of Chelyabinsk were chosen as a standard. Information on radiation exposure levels of thyroid for Ozyorsk residents of different ages was obtained on the basis of preliminary dose calculations for two scenarios of food consumption from different suppliers. Radiation excess relative risk per unit dose (ERR/Sv) was calculated using a Poisson regression model implemented in Amfit module of the Epicure software.

Results: SIR of thyroid cancer in the cohort was 3.16 among men (1.68–5.06) and 2.07 (1.52–2.65) among women compared to the data of national statistics. Significant difference of thyroid cancer incidence in the cohort compared to the regional statistics was also found: SIR was 2.04 (1.08–3.26) among men and 1.59 (1.17–2.04) among women. ERR/Sv was –0.03 ($p>0.5$), i.e. increase of thyroid cancer incidence in the study cohort is not associated with the obtained estimates of equivalent doses to the thyroid.

Conclusion: In the cohort of persons residing as children in Ozyorsk in the period of uncontrolled gas-aerosol emissions from the chimneys of radiochemical plants in 1948–1962, age- and sex-related thyroid cancer incidence rates and relative risk of thyroid cancer incidence in men and women significantly (2–3 times) exceeded those in the national and regional statistics. The correlation between dose and effect was not observed. It could be associated with using group doses due to lack of individual doses.

References

1. L.A. Ilyin et al. *Radioactive iodine in radiation safety problem*. (Atomizdat, Moscow, 1972).
2. A.V. Akleyev, L.U. Krestinina, D. Preston, Et al. Med. radiol. and rad. safety, **53**, P.13–37 (2008).
3. U.V. Glagolenko, E.G. Drozhko, U.G. Mokrov, et al. Radiation safety issues. Special volume, P.35–51 (2008).
4. U.V. Glagolenko, E.G. Drozhko, U.G. Mokrov, et al. Radiation safety issues. Special volume, P. 52–61 (2008).
5. L.A. Ilyin, E.M. Aksel, E.G. Drozhko et al. Med. radiol. and rad. safety, **48**, P.57–64 (2003).

Assessment of lifetime attributable risks from internal radiation exposure

Aleksandr Menyajlo[1], *Sergey* Chekin[1], *Valery* Kashcheev[1], *Marat* Maksioutov[1], *Aleksandr* Korelo[1], *Konstantin* Tumanov[1], *Evgeny* Pryakhin[1], *Sergey* Lovachev[1], *Sergey* Karpenko[1], *Polina* Kashcheeva[1], and *Viktor* Ivanov[1].

[1]Federal State Budgetary Institution National Medical Research Radiological Center of the Ministry of Health of the Russian Federation

The method for estimating lifetime attributable risk from internal exposure to radiation with account of the sex and age of an exposed person is presented. Initially it was developed to implement the principle of socially acceptable risk in Russia. By the Russian radiation safety standards the acceptable individual lifetime risk from planned exposure should not exceed 5×10^{-5} and 1×10^{-3} for the members of public and workers respectively. These limits were standardized using ICRP nominal risk coefficients, calculated for a composite population and cannot be used for making prediction of effects of radiation exposure on a real population. Therefore, it is reasonable to use the models recommended by ICRP for estimating radiation risks.

The calculation of radiation risks from internal exposure is based on the modern ICRP models [1] and on the dynamics of equivalent doses in various organs and tissues. The dose coefficients (Sv/Bq) allow to transfer the activity of different radioisotopes in organs and tissues to equivalent radiation doses depending on the time after intake. This information is taken from the ICRP database [2]. The database contains dose coefficients for 738 different radionuclides. Data are available for both inhalation and ingestion radionuclides. For inhalation intake – for each radionuclide, there are dose coefficients for 10 different values of the median aerodynamic diameter (AMAD) activity, μm: (0,001, 0,003, 0,01, 0,03, 0,1, 0,3, 1,0, 3,0, 5,0, 10,0) and three solubility options: slow (S), intermediate (M) and fast (F). In addition, the database takes into account the age at intake of radionuclides. Also, there are separate coefficients for workers and members of public.

The presented method for estimating lifetime attributable risk developed in accordance with international recommendations is the scientific support to the principle of socially acceptable radiation risk. Risk assessments are used for predicting radiation effects on human health as a result of planned, existing and emergency exposure. The method can be useful for experts in radiological protection and radiation hygiene, referring medical practitioners and radiologists, medical physicists and radiobiologists.

References

1. ICRP Publication 103. Eds.: M.F. Kiselev, N.K. Shandala. Moscow, PKF «Alana», 2009. 312 p. Available at: http://www.icrp.org/docs/P103_Russian.pdf (Accessed 16.05.2018). (In Russian).

2. ICRP Database of Dose Coefficients: Workers and Members of the Public; Ver. 3.0, official website. Available at: http://www.icrp.org/page.asp?id=145 (Accessed 16.05.2018).

The United States Transuranium and Uranium Registries: Fifty-year History of Actinide Biokinetic Research

Maia Avtandilashvili[1*], *Stacey L.* McComish[1], and *Sergei Y.* Tolmachev[1]

[1] U.S. Transuranium and Uranium Registries, College of Pharmacy and Pharmaceutical Sciences, Washington State University, 1845 Terminal Drive, Suite 201, Richland 99354, WA, USA

Introduction

Designed as a program to improve radiation protection of nuclear workers, the U.S. Transuranium and Uranium Registries (USTUR) studies the biokinetics and internal dosimetry of actinides (uranium, plutonium, and americium) in occupationally exposed Registrants who volunteer portions of their bodies, or their whole bodies, for scientific use posthumously [1]. The USTUR is an invaluable national and international resource for testing biokinetic models and improving the application of bioassay data to estimate tissue doses.

Established as the National Plutonium Registry (NPR) in 1968 by the U.S. Atomic Energy Commission (AEC), today, the USTUR is one of the longest-operating research programs funded by the U.S. Department of Energy (DOE). Since 1992, the Registries is a grant-funded program operated by College of Pharmacy and Pharmaceutical Sciences at Washington State University (Fig 1).

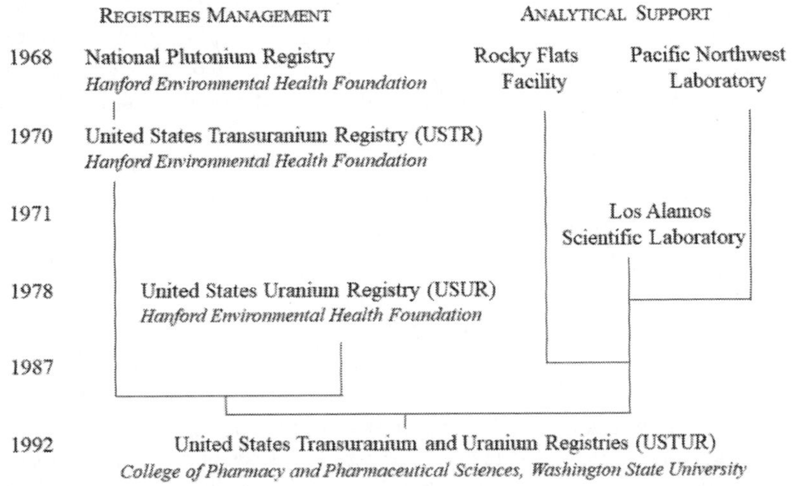

Fig. 1. Genealogy of the USTUR.

---

* Corresponding author: m.avtandilashvili@wsu.edu

USTUR Registrants

Currently, the USTUR retains records and data for 310 partial-body donors, 46 whole-body donors, and 33 living Registrants. These individuals typically worked at defence nuclear facilities where plutonium, americium, or uranium were processed (Fig 2).

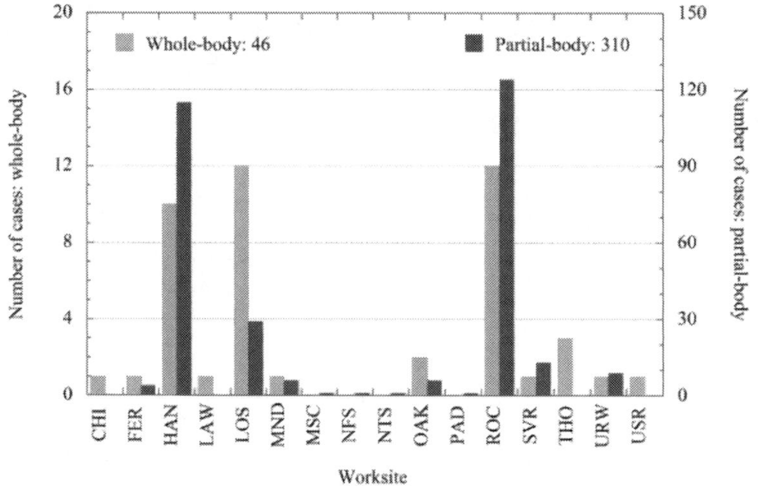

Fig. 2. USTUR Registrants by worksites.

Historical Landmarks

First whole-body donation: USTUR Case 0102. The first whole-body donation to the Registries was in 1979. This individual had a known exposure to ^{241}Am [2]. Half of the skeleton was analyzed, and the remaining half was used to make *in vivo* calibration phantoms of skull, torso, arm and leg, which have been made available to other scientists and are used worldwide [3,4].

The Atomic Man: USTUR Case 0246. This partial-body donor is known to have had the largest recorded intake of ^{241}Am [5]. An explosion of an ion-exchange column resulted in an estimated uptake of 40 MBq. He was heavily chelated with Ca/Zn-DTPA. This treatment reduced systemic deposition of ^{241}Am to 0.54 MBq and is believed to have saved his life.

Thorotrast study: USTUR Case 1001. This whole-body donor was medically exposed to Thorotrast®. Post-mortem analyses of tissues from this individual provided a unique opportunity to study the distribution and biokinetics of thorium [6].

These three USTUR cases were described in detail in four special issues of the Health Physics journal.

Research

The uniqueness of the USTUR lies in its ability to link thoroughly documented exposure, work history, medical, industrial hygiene, and bioassay data with precise post-mortem measurements of the content and distribution of radionuclides in the human body. Historically, the research at the USTUR has focused on studying long-term retention and distribution of uranium, thorium, plutonium, and americium in the human body and biokinetic modelling of these elements. Post-mortem radiochemical analyses of tissues obtained at autopsy, especially those from the whole-body donors, allow the USTUR to significantly improve our knowledge of the distribution and long-term retention of actinides in the human body and have helped to parameterize biokinetic constants for these radioactive elements. During the last decade, the USTUR research was expanded to include modelling of actinide decorporation therapy and evaluation of uncertainties in organ dose estimates. Most recently, the USTUR explored opportunities to study beryllium and other non-radioactive materials in the human body. A few examples of recent USTUR research include evaluation of long-term plutonium retention in the upper airways to quantify plutonium 'bound' fraction [7], modelling of enriched uranium hexafluoride inhalation [8], and development of a system of compartmental models for plutonium decorporation [9].

The USTUR has published over 300 scientific papers. The Registries' research findings have contributed to five reports published by the National Council on Radiation Protection and Measurements (NCRP) and eight International Commission on Radiological Protection (ICRP) publications including the upcoming Occupational Intakes of Radionuclides, Part 4. A special issue of the Health Physics journal dedicated to USTUR research is currently in production.

Conclusions

The year 2018 marks 50 years of research at the U.S. Transuranium and Uranium Registries. The USTUR is the only program worldwide that can study biokinetics and dosimetry of internally deposited actinides this comprehensively. The USTUR serves as a source for both scientific research and public information regarding the radiobiology, biokinetics and dosimetry of the actinide elements in humans. Clearly, the USTUR's accomplishments could happen only because of our Registrants – without whom nothing would be possible.

References

1. The United States Transuranium and Uranium Registries. https://ustur.wsu.edu/
2. G. R. Roessler. Health Phys. **49**, 559 (1985).
3. G. H. Kramer, B. Hauck, K. Capello et al. Health Phys. **101**, 248 (2011).
4. P. Nogueira, W. Ruhm, M.A. Lopez et al. Radiation Meas. **82**, 64 (2015).
5. R. E. Toohey, R. L. Kathren. Health Phys. **69**, 310 (1995).
6. R. L. Kathren, R. L. Hill. Health Phys. **63**, 72 (1992).
7. M. Puncher, A. Birchall, S. Y. Tolmachev. Radiat. Prot. Dosim. **176**, 50 (2017).
8. M. Avtandilashvili, M. Puncher, S. L. McComish, S. Y. Tolmachev. J. Radiol. Prot. **35**, 129 (2015).
9. S. Dumit, M. Avtandilashvili, S. L. McComish, D. J. Strom, G. Tabatadze, S. Y. Tolmachev. Radiat. Res. (submitted).

Effect of natural uranium on bone cells

Tatiana Gritsaenko[1], *Lucile* Hurault[1], *Valérie* Pierrefite-Carle[1], *Sabine* Santucci-Darmanin[1], *Sophie* Pagnotta[2], *Claude* Vidaud[3], *Agnès* Hagège[4], *Pier Lorenzo* Solari[5], *Gaëlle* Creff[6], *Christophe* Den Auwer[6], and *Georges F.* Carle[1]

[1]UMR E-4320 TIRO-MATOs CEA/DRF/BIAM, Université Nice Sophia Antipolis, Nice, France
[2] Centre Commun de Microscopie Appliquée, Université Nice Sophia Antipolis, Nice, France
[3] CEA/DRF/BIAM, Bagnols sur Cèze, France
[4]Institut des Sciences Analytiques, UMR 5280 CNRS, Villeurbanne, France
[5]Synchrotron SOLEIL, Gif-Sur-Yvette, France
[6]Institut de Chimie de Nice, UMR 7272 CNRS/UNS, Université Nice Sophia Antipolis, Nice, France

Natural uranium (NU), a mix at equilibrium of U-238, U-235 and U-234, is a natural element of the earth crust presenting a dual toxicity related both to its radiological effects as an alpha emitter and its chemical effects due to its metals properties. In atmospheric conditions and in most environmental systems, uranium is found in its oxocationic form $\{U(VI)O^{2+}{}_2\}$ that is the most stable form of uranium at formal oxidation state +VI simply referred as "U(VI)" hereafter.
Although many studies have investigated the effects of acute U(VI) exposure on living organisms, the consequences of chronic exposure to low doses on various cell fonctions remain elusive. The large majority of ingested U(VI) is rapidly excreted, however once it reaches the blood compartment, close to 20% will end up in skeletal tissue considered as the long term storage organ.
Considering the low specific activity of NU, we have adressed the chemical toxicity of this actinide on the viability, and function of the bone cells in charge of bone turnover: osteoblasts (OB), the cells in charge of bone matrix synthesis and mineralisation, osteoclasts (OC), the ones taking care of bone degradation and resorption, and osteocytes (OST), the more abundant ones, acting as mechanosensors and orchestrators of bone remodeling .
Using a rodent osteoblastic cell line (UMR-106), we have shown that the presence of U(VI) at sub-toxic concentrations as low as 25 μM, alters the matrix mineralization function. In comparison, investigating the effect of U(VI) of a mouse osteocyte cell line (MLO-A5), we observed that although the cytotoxicity index was more than twice higher than for the OB cell line, the mineralization function of these cells was affected starting at 2.5 μM of U(VI). Moreover, when exposed to sub-toxic doses of U(VI) for 24hrs, we could observe in both models the presence of needle like structures outside the cells as well as within vesicles inside the cells by Transmission Electron Microscopy. In order to determine the speciation of uranium in these models, EXAFS spectroscopic analysis was undertaken and revealed the presence of autunite forms of this actinide in both OB and OST cell types.
The involvment of autophagy, a self cleaning and recycling process, also shown to be crucial for mineralization of OB, has been investigated upon U(VI) exposure in OB and OST and will be discussed.
Finally, we examined the consequences of U(VI) exposure on both a pre-osteoclastic cell line (RAW264.7) as well as primary murine osteoclastic cells. We observed that when U(VI) is present in solution at concentrations in the μM range, it impairs OC formation,

mature OC survival, as well as mineral resorption function. We are presently investigating the effect of U(VI) imbedded in the matrix on the same processes.

Altogether, these results show that NU can affect the major bone cell types in their viability, differenciation and functions even at sub-toxic doses thereby impacting bone homeostasis.

BIO Web of Conferences **14**, 05003 (2019) https://doi.org/10.1051/bioconf/20191405003
HEIR 2018

Effect of radiotoxic effect of uranium ore dust on the reproductive system of rats

Sevda Toklayeva[1], *Yelena* Saifulina[1], *Saltanat* Utebayeva[1], *Ayagoz* Ibrayeva[1], *Meirat* Bakhtin[1], *Polat* Kazymbet[1]

[1]Institute of radiobiology and protection JSC Astana medical university

The purpose of research was to study the reproductive function of female and male rates when exposed to uranium ore dust.
A series of studies was performed on 43 mature white mongrel rats of both sexes. Oral intake of UOD was simulated for 7 days of daily feeding of 30 white outbred rats with uranium ore dust (UOD) mixed with feed. The mass of uranium ore dust for single feeding of animals of the experimental group in a dose of 50 is the maximum allowable concentration (MAC) was 5750 mg.

The control group consisted of intact white rats with normal feeding. 24 hours after the last feeding of (UOD), some animals reproduced biomaterials of the reproductive system of females and males and measured the concentration of uranium in samples on an inductively coupled plasma mass spectrometer "Agilent-7800" with a range of measured masses from 2 amu up to 260 amu.

After irradiation, the experimental animals were divided into 3 subgroups for crossing: I was represented by irradiated females (n = 12) and intact males, II subgroup by intact females (n = 7) and irradiated males, and III - control subgroup of males and females with normal diet. Experimental and control female rats were planted with males in a ratio of 5: 2.

According to the results of our study in the organs of the reproductive system of the female uterus and ovaries, the uranium content was more than 2 times higher than that in the control animals, and in the testes of males more than 1.5 times. The sexual function was judged based on the study of crossbreeding and compared it with the control group, where the onset of pregnancy was observed in 100% of cases. The state of sexual and reproductive function of experimental rats along the course of pregnancy and childbirth. The gestation period of pregnancy in irradiated animals lasted an average of 26-27 days in comparison with control females. Of the female rats of the experimental group who were mated with intact males, 58% became pregnant, a similar pattern was observed in the group of irradiated males crossed with intact females - 57%, that is, the cases of pregnancy in rats in both groups decreased almost 2 times compared to the control group .

The result of the studies indicates that the radiotoxic effects of rats result in a decrease in the ability to reproduce both females and males.

In vivo comparison between two nephrotoxic agents, sodium fluoride and uranyl nitrate: phenotypic aspects and molecular mechanisms involved

Alice Bontemps[1,*], *Laurine* Conquet[1], *Christelle* Elie[1], *Victor* Magneron[1], *Céline* Gloaguen[1], *Dimitri* Kereselidze[1], *Karine* Tack[1], *Olivier* Barbier[2], and *Yann* Guéguen[1]

[1]Institut de Radioprotection et de Sûreté Nucléaire (IRSN), PRP-HOM, SESANE, LRTOX, 92262 Fontenay-aux-Roses, France
[2]Departamento de Toxicología, Centro de Investigación y de Estudios Avanzados del Instituto Politécnico Nacional (CINVESTAV-IPN), Av. IPN No. 2508 Col, San Pedro Zacatenco, México City, CP 07360, Mexico

1 Background

Humans can be exposed at low concentrations to toxic agents with anthropogenic and natural origins such as uranium and fluoride. Because of its functions of filtration, transport and reabsorption the kidney is a primary target organ of toxicity to foreign compounds. Uranium and fluoride are both known to be nephrotoxic, nevertheless there is a lack of knowledge of their mechanisms of nephrotoxicity and of the underlying molecular pathways involved [1, 2]. This study aims to compare these agents to identify the cellular and molecular pathways of nephrotoxicity in mouse.

2 Methods and results

C57Bl6 mice are exposed to uranyl nitrate (UN) (0, 2, 4, 5 mg/kg) or sodium fluoride (NaF) (0, 2, 5, 7.5, 10 mg/kg) by intraperitoneal injection and put into metabolism cages for 24 hours before and directly after injection. Collected urines are used to quantify uranium or fluoride, to measure specific and sensitive biomarkers of nephrotoxicity levels and to assess clinical bioassays. Finally, animals are euthanized 48h and 72h after exposure, chosen as the peak of nephrotoxicity is observed after intraperitoneal injection [3]. The protocol is shown in figure 1.

Renal phenotypic aspects and biological mechanisms are evaluated by urinary biochemistry, gene and protein expressions, enzyme activity, and histological analyses. Exposure to UN and NaF induces nephrotoxicity in a dose-dependent manner. A 5 mg/kg injection of UN induces mild histopathological alterations and respectively 44 and 6-fold increase in gene expressions of nephrotoxicity markers KIM1 and osteopontin. In comparison, 10 mg/kg of NaF induces high nephrotoxicity with histopathological alterations scored as severe and late appearing parameters of toxicity whereas 7.5 mg/kg

* Corresponding author: alice.bontemps@irsn.fr

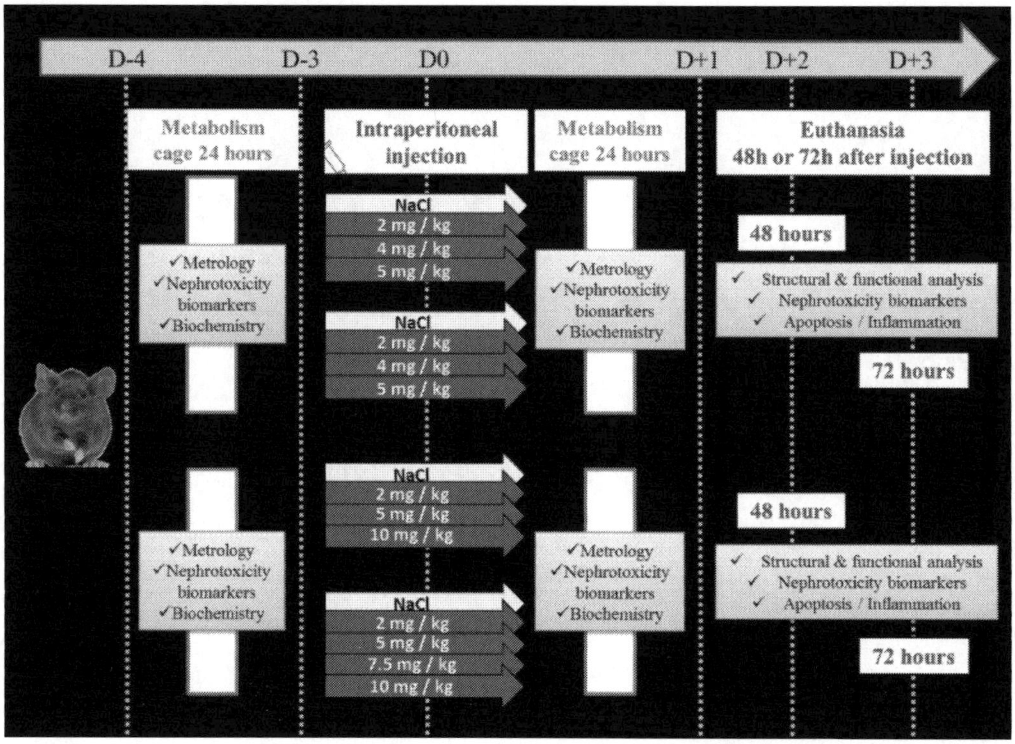

Fig. 1. Protocol of internal exposure of uranium and fluoride in mice C57Bl6. Scale of time is given at the top of the figure (in days). Day 4 before injection mice are put 24 hours in metabolism cages to collect urines, at day 0 mice are injected intraperitoneally with either uranium or fluoride and directly put in metabolism cages again. Day 2 or day 3 after injection mice are euthanized to extract kidneys.

induces mild histopathological scoring and gene expressions of KIM1 and clusterin enhanced respectively by 70 and 4-fold compared to control. No signs of nephrotoxicity are observed below 5 mg/kg of NaF. Clusterin is respectively increased by 2.4 and 4.4-fold in urines after 7.5 and 10 mg/kg injection of NaF, whereas 4 and 5 mg/kg of UN induce respectively a 2.2 and 2.7-fold increase of clusterin in urines. Apoptosis is evaluated through caspases 3/7 activity which is increased by 210% after UN treatment (5mg/kg) whereas NaF does not induce apoptosis significantly. Inflammation is implied in UN and NaF acute nephrotoxicity as shown by gene and *in situ* overexpressions of ICAM and VCAM measurement by immunohistochemical staining.

3 Conclusions

UN and NaF acute exposures resulted in dose and time-dependent nephrotoxicity with a higher nephrotoxicity after 72h. Inflammation and apoptosis are both involved in UN or NaF toxicity. These observations allow us to identify the mechanisms that will be studied in a low-dose exposure protocol after a chronic exposure.

References

1. Y. Gueguen, & C. Rouas, Radioprotection, 47(3), 345-359 (2012)
2. O. Barbier, L. Arreola-Mendoza, et al., Chem Biol Interact 188(2): 319-333 (2010)
3. P. Sangeetha Vijayan, P. D. Rekha, et al., Ren Fail 38(5): 770-775 (2016)

BIO Web of Conferences **14**, 05005 (2019) https://doi.org/10.1051/bioconf/20191405005
HEIR 2018

Radon induced hyperplasia may provide an explanation for inverse exposure rate effect

Balázs Madas*, and *Emese* Drozsdik

MTA Centre for Energy Research, Environmental Physics Department, Konkoly-Thege Miklós út 29-33, 1121 Budapest, Hungary

1 Introduction

In most cases, incorporation of radionuclides results in protracted exposures. It means that adaptive responses may affect the health consequences more significantly than in case of acute exposures. It is reasonable to assume that a given dose results in lower risk if the dose rate is lower and the duration is longer, as it is reflected in the application of the dose and dose rate effectiveness factor (DDREF) [1]. In case of inhaled radon progeny, however, there is evidence for an inverse exposure rate effect [2] meaning that longer duration and lower exposure rate results in higher risk for a given exposure.

Inhalation of radon progeny is characterized by highly heterogeneous dose distribution within the bronchial airways [3]. While the spatial and temporal distribution of dose does not modulate biological effects which show linear dose dependence, they may have significant impact in case of non-linear biological responses to ionizing radiation such as radioadaptive responses. Histological and experimental studies show that different cell types increase in number in response to a wide variety of drugs and irritants [4,5] resulting in hyperplasia in the bronchial epithelium. Although there is no direct evidence for the induction of hyperplasia upon radon exposure, theoretical considerations about its effectiveness in adaptation to extensive cell loss [6] supports the idea that hyperplasia occurs in the deposition hot spots upon chronic exposures to radon progeny, as well.

The objectives of the present study are to quantify how the induction of hyperplasia modulates the microdosimetric consequences of a given (macroscopic) exposure and to discuss whether the phenomenon can provide a link between two characteristics of radon exposure: heterogeneity in dose distribution and inverse exposure rate effect.

2 Methods

Based on histological data [7,8], computational epithelium models with different thickness and different abundance of basal and goblet cells were prepared to represent different measures of hyperplasia. Spherical cell nuclei of six different cell types were considered. As goblet cells play a major role in mucus production, the thickness of mucus layer covering the epithelium was supposed to be proportional to goblet cells number. Applying Monte-Carlo codes, cellular burdens such as cell nucleus hits and cell nucleus doses were quantified for the deposition hot spots, where local doses can be two orders of magnitude

* Corresponding author: balazs.madas@energia.mta.hu

higher than the average over the airway generation [3]. Besides average cell nucleus hits and dose, distribution of hit numbers and doses of cell nuclei were determined.

3 Results and discussion

Results show that average tissue dose, average hit number and dose of basal cells decrease by the increase of the measure of both basal and goblet cell hyperplasia [9,10]. Hit and dose distributions reveal that the induction of hyperplasia may result in a basal cell pool which is shielded from alpha-particles in particular in case of goblet cell hyperplasia where both mucus thickening and the additional number of goblet cells decrease cellular burdens.

The induction of hyperplasia points out that the exposure history affects the microdosimetric consequences of a present exposure. It can be considered as a radioadaptive response at the tissue level which challenges the validity of the application of the dose and dose rate effectiveness factor from a mechanistic point of view.

As the location of radiosensitive target cells may change due to previous exposures, dosimetry models considering the tissue geometry characteristic of normal conditions may be inappropriate for dose estimation in case of protracted exposures. As internal exposures are frequently chronic, such changes in tissue geometry may be highly relevant for other incorporated radionuclides too.

Based on theoretical considerations [6], it is expected that the measure of hyperplasia increases by exposure rate. As the increased measure of hyperplasia results in lower dose rates, a given total exposure can cause more genetic damage in the tissue if the duration is longer and the exposure rate is smaller. In this way, the induction of hyperplasia can provide an explanation for inverse exposure rate effect. In addition, this kind of adaptation of the tissues may also play a role in the observation that heterogeneous exposures result in lower risk compared to homogeneous exposures as it is seen in case of hot particles [11].

This work was supported by the National Research, Development and Innovation Office under the contract VKSZ_14-1-2015-0021.

4 References

[1] International Commission on Radiological Protection, *Ann ICRP,* **37** (2007) 1–332.
[2] J.H. Lubin, J.D. Boice, C. Edling, R.W. Hornung, G. Howe, W. Kunz, R.A. Kusiak, H.I. Morrison, E.P. Radford, J.M. Samet, M. Tirmarche, A. Woodward, S.X. Yao, *Health Phys,* **69** (1995) 494–500.
[3] I. Balásházy, Á. Farkas, B.G. Madas, W. Hofmann, *J Radiol Prot,* **29** (2009) 147–162.
[4] D.F. Rogers, *Int J Biochem Cell Biol* **35** (2003) 1–6.
[5] I.O. Gordon, S. Sitterding, A.C. Mackinnon, A.N. Husain, *Arch Pathol Lab Med,* **133** (2009) 1106–1115.
[6] A.D. Lander, K.K. Gokoffski, F.Y.M. Wan, Q. Nie, A.L. Calof, *PLoS Biol,* **7** (2009) e15.
[7] R.R. Mercer, M.L. Russell, J.D. Crapo, *Health Phys,* **61** (1991) 117–130.
[8] R.R. Mercer, M.L. Russell, V.L. Roggli, J.D. Crapo, *Am J Respir Cell Mol Biol,* **10** (1994) 613–624.
[9] B.G. Madas, *J Radiol Prot,* **36** (2016) 653–666.
[10] B.G. Madas, E.J. Drozsdik, *Int J Radiat Biol,* **94** (2018) 967–974.
[11] M.W. Charles, A.J. Mill, P.J. Darley, *J Radiol Prot,* **23** (2003) 5–28.

BIO Web of Conferences **14**, 05006 (2019) https://doi.org/10.1051/bioconf/20191405006
HEIR 2018

Using mRNA and small RNA gene expression changes in peripheral blood for easy detection of Ra-223 incorporation

Patrick Ostheim[1,*], *Matthäus* Majewski[1], *Matthias* Miederer[2], *Mathias* Schreckenberger[2], *Tim* Nestler[3], *Manuela A.* Hoffmann[4], *Matthias* Port[1], and *Michael* Abend[1]

[1] Bundeswehr Institute of Radiobiology, Munich, Germany
[2] University Medical Center of the Johannes Gutenberg University, Clinic and Polyclinic for Nuclear Medicine, Mainz, Germany
[3] Department of Urology, Federal Armed Services Hospital Koblenz, Koblenz, Germany
[4] Supervisory Center for Medical Radiation Protection, Bundeswehr Medical Service Headquarters, Koblenz, Germany

1 Background

Radiation analytics is the established method for detection of incorporated alpha-emitting radionuclides. It is laborious and time consuming. We wondered whether changes in gene expression measured in the human peripheral blood after radionuclide incorporation might serve as an easier approach for detection of incorporated radionuclides in a radiological or nuclear scenario.

2 Methods and Results

We received whole blood in PAXgene Blood RNA tubes from four patients suffering from multiple bone-metastatic, castration-resistant prostate cancer (without visceral or nodal involvement), who underwent treatment with the alpha emitting isotope radium-223 dichloride (Ra-223, Xofigo®). Patients received about 4 MBq per cycle and month and were treated for six months.

In a first step, we employed next generation sequencing (NGS) for a whole genome screening of the transcriptome (mRNAs) as well as the post-transcriptome including small RNAs (e.g. long non-coding RNAs, snoRNAs and miRNAs) in one patient at 8 different time points during 6 cycles of Ra-223-therapy. This allowed us to identify specific candidate RNAs which are associated with the increased dose cumulated over the treatment time.

*Corresponding author: patrickostheim@bundeswehr.org

In a second step, candidate mRNAs and small RNAs which appeared dose-dependently differentially up- or down-regulated (>2-fold, $p<0.05$) in the screening samples will be chosen for validation. This will be performed by using the remaining patient samples ($n=3$ patients, each with up to 4 points in time) and shifting to more sensitive and specific qRT-PCR platform at the same time.

3 Conclusion/Prospect

We will present and discuss the results on radiation-induced gene expression changes in whole blood which might serve as a simplified diagnostic tool for identification of incorporated radionuclides (Ra-223) in future.

Key words
Radiation analysis, incorporation, gene expression changes, radiation biomarkers, small RNA, mRNA, transcriptome, prostate cancer, Radium-223.

Effect of repetitive potassium iodide on elderly rat's thyroid

Dalila Lebsir[1, *], *Elsa* Cantabella[1], *Teni* Ebrahimian[1], *Dimitri* Kereselidze[1], *Stephane* Grison[1], *Amandine* Sache[1], *Karine* Tack[1], *Marc* Banderitter[1], *Annick* Pech[2], *Philippe* Lestaevel[1], *Maâmar* Souidi[1]

[1]Institut de radioprotection et de sureté nucléaire, 92262 Fontenay-aux-Roses, France
[2]Pharmacie centrale des armées, 45000 Orléans, France

Abstract

Background: Nuclear power plant emergencies had often been accompanied by radioactivity release into the environment, thyroid cancer is one of the major health consequences due to the effect of radioactive iodine (^{131}I) that emits ϒ ray and β particles resulting in thyroid DNA damage and late onset thyroid cancer. Intake of a single dose of potassium iodide (KI) is recommended to reduce this risk. However in case of prolonged radioiodine release as noticed during Chernobyl and Fukushima accidents, more than one dose of KI may be basic to ensure adequate protection [1]. Whereas a single dose of KI is admitted to be safe, knowledge about the effects of repeated KI administration are scarce, few studies demonstrated the potential efficiency of repetitive KI intake in humans [2] and non-human primates [3] without hormonal variations. These studies are relevant in the field of radiation protection and give a base evidence of the possible use of repetitive KI. On the other hand, we have studies on rodents that showed an impact of chronic iodine excess on pituitary thyroid axis function [4]. Our previous work on adult male rats demonstrated the safety of repeated administration of KI over 8 days [5]. Indeed in the elderly persons KI administration in case of nuclear emergency remains a topic of debate, because of the possible impact in cardiovascular diseases. Thyroid hormones are well-known for their profound effects on cardiovascular function and metabolism; myocardial and vascular endothelial tissues have receptors for thyroid hormones and are sensitive even to subtle changes in the concentrations of circulating pituitary and/or thyroid hormones i.e. subclinical hypothyroidism and hyperthyroidism. It is well established that hyperthyroidism induces a hyper-dynamic cardiovascular state, which is associated with a faster heart rate, enhanced left ventricular systolic and diastolic function whereas hypothyroidism is characterized by the opposite changes. Atrial fibrillation is the most common cardiac arrhythmia in the elderly, the prevalence and incidence increase with advancing age [6]. Several interventional trials showed that treatment of subclinical thyroid diseases improves cardiovascular risk factors, which implies potential benefits for reducing cardiovascular events. **Objective**: The aim of this study is to assess the effects of repeated KI intake on the thyroid function of aged male rats. **Methods**: A twelve months old male rats were subjected to either KI or saline solution over 8 days. Clinical biochemistry, pituitary and

* Corresponding author: dalila.lebsir@irsn.fr

thyroid hormones level, and thyroid genes expression were analyzed 30 days after the treatment discontinuation. **Findings**: urinary assessment shows a subtle increase of some parameters (Creatinin, Uric Acid, Urea, Glucose, Potassium, Sodium and Chlorine), plasma biochemistry reveals a subtle variation of some parameters (an increase of Creatinin, Glucose and phosphorus; and a decrease of Chlorine level). Regarding pituitary-thyroid hormones we get a significant decrease of TSH level without thyroid hormones variation. At the molecular level, we observe a significant increase of TPO (+100%), AIT (+299%) and Tg (+38%) mRNA expression. On the other hand we get a significant decrease of TSHR (-51%) mRNA expression. **Conclusion and perspectives:** Our first results indicate that repeated KI intake affects the clinical biochemistry and the pituitary-thyroid axis function in elderly rats. To go further we are investigating the impact of these variations on the cardiovascular function and its parameters. Cardiac output data, cardiovascular gene expression, oxidative stress and inflammatory analysis are being processed. This study will contribute to the evolution of iodine policy and the harmonization of the current KI guidelines.

Figure

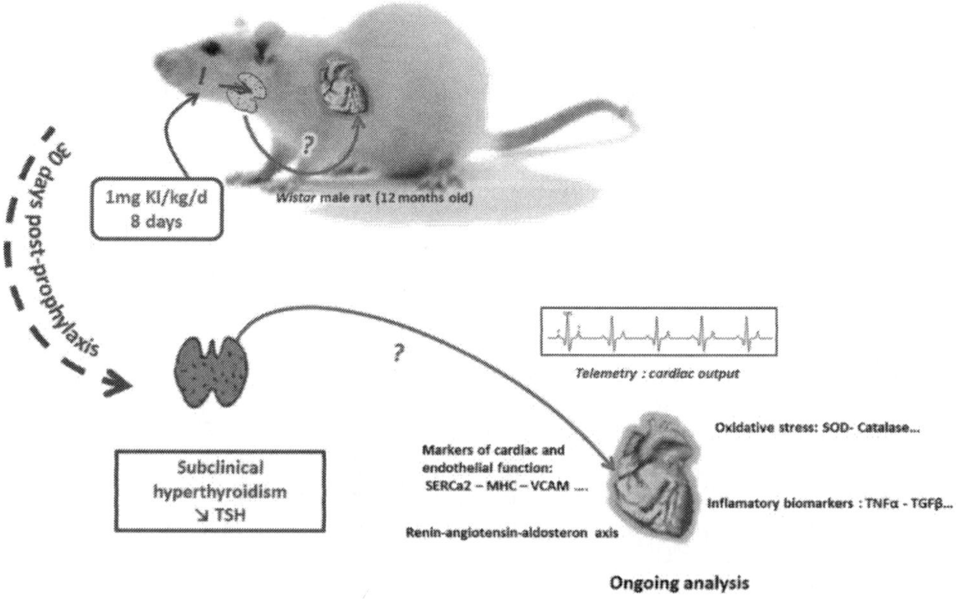

Fig. 1. Impact of repeated KI intake on the thyroid of elderly male rat

References

1. WHO. ISBN 978 92 4 1550185 (2017).
2. Sternthal, E., et al. " N Engl J Med 303(19): 1083-1088. (1980).
3. Noteboom JL., et al. Radiat Res.147(6):698-706. (1997).
4. Calil Silveira J., et al. Thyroid. 26:26. (2016).
5. Lebsir D., et al. Mol Cell Endocrinol.26(18):30077-7. (2018).
6. Delitala AP. Horm Metab Res.49(10):723-31. (2017).

A putative mechanism of the Sodium/Iodide Symporter regulation during repetitive administration of stable Iodide described by a Systems Biology approach

David Cohen[1], *Dalila* Lebsir[1], *Karine* Tack[1], *Marc* Benderitter[2], and *Maâmar* Souidi[1]

[1] Institut de Radioprotection et de Sûreté Nucléaire (IRSN), PSE-Santé/SESANE/LRTOX, 92262 Fontenay-aux-Roses, France
[2] Institut de Radioprotection et de Sûreté Nucléaire (IRSN), PSE-Santé/SERAMED, 92262 Fontenay-aux-Roses, France

Abstract

A single dose of potassium iodide (KI) against a prolonged exposure to repeated radioactivity might not be effective enough to protect the thyroid. Our group have shown that a repetitive dose of KI for eight days offers efficient protection without adverse effects in male rats [1].

However, we also have shown that the expression of the genes involved in the Wolff-Chaikoff effect changes during this period. Notably, a decrease in the sodium/iodide symporter (*nis*) gene expression has been observed [1]. This effect may result in hypothyroidism due to a decrease in thyroid hormones.

NIS is responsible for the uptake of KI and thus plays an important role in the Wolff-Chaikoff effect. The mechanism of a single dose of KI on the toxicity of the thyroid is well known [2], in contrast to repetitive administration of KI for eight days.

In the present study, we try to understand the Wolff-Chaikoff regulation and its molecular constituents during repetitive administration of KI. For this purpose, we have constructed manually a biochemical reaction network that is visualised as a "geographical" map of a single thyrocyte cell depicting the iodide and thyroid hormone metabolism. In order to investigate any regulation circuits of *nis*, Cytoscape and the plugin BiNoM [3, 4] were used to perform path analysis of the network to investigate if a path exists from the node iodide going to the node representing "nis transcription". Subsequently, sequential network reduction has led to final model that might explain a putative mechanism behind *nis* regulation and repetitive iodide administration.

In addition, this map reviews the most-update information about iodide and thyroid hormone metabolism. Besides as a source of information, it can help to elucidate the mode of action of KI on gene transcription after repetitive KI administration.

References

1. D. Lebsir, L. Manens, S. Grison, P. Lestaevel, T. Ebrahimian, D. Suhard, G. Phan, I. Dublineau, K. Tack, M. Benderitter, A. Pech, J.-R. Jourdain, M. Souidi. Mol. Cell. Endocrinol. (2018), doi: 10.1016/j.mce.2018.02.017

2. D. Carvalho, C. Dupuy. Mol. Cell. Endocrinol. **458,** 6 (2017)
3. P. Shannon, A. Markiel, O. Ozier, N.S. Baliga, J.T. Wang, D. Ramage, N. Amin, B. Schwikowski, T. Ideker. Genome Res. **13** 2498 (2003)
4. E. Bonnet, L. Calzone, D. Rovera, G. Stoll, E. Barillot, A. Zinovyev. BMC Syst Biol. **1** 7 (2013)

Histological long-term effects in relevant tissues after 131I contamination of two Wistar rat models

Meriem Mezaguer-Lekouaghet[1,*], *Abdelwahab Badreddine*[1], *Saadia Mameri*[2], *Maamar Souidi*[3], *Ahcène Baz*[4], *and Zohra Lounis-Mokrani*[1]

[1] Centre de Recherche Nucléaire d'Alger, Algérie
[2] Centre Hospitalo Universitaire Mustapha Bacha Alger, Algérie
[3] Institut de Radioprotection et de Sûreté Nucléaire Paris, France
[4] Ecole Normale Supérieure de Kouba, Algérie

Abstract

Procedures using iodine-131 (^{131}I) represent about 90% of all therapies in nuclear medicine [1, 2]. It is important to evaluate the long-term biological effects of ^{131}I treatment on non-target organs in order to improve the patient radioprotection.

The aim of this work is to use an experimental animal model to contribute to the understanding of some biological effects induced by ^{131}I contamination, and especially histological effects.

Two models of Wistar rats with (Tr+) and without thyroid (Tr-) were orally contaminated with iodine and after 8 months they were sacrificed and the histological effects on some relevant tissues such as thyroid, testes, liver and kidneys were been studied.

Thyroid tissue analysis revealed that in the case of the Tr+ model compared to their control (Tr+ uncontaminated), 50% of the slides examined had remodeled the thyroid tissue with rare follicles choked by fibrosis and with epithelial changes. However, for Tr- compared to their control, the examined slides reveal the presence of a small, completely atrophied thyroid mass associated with vesicular fibrosis and with detachment of the colloid.

For the renal organ, disturbances are observed: inflammation of the presence of tissue fibrosis and glomerular necrosis. For the liver, there is an appearance of inflammatory focus in different degrees around the portals. However, the results of the testes of both models compared to their controls revealed no histological abnormalities. The observed histological effects are correlated with the corresponding absorbed dose received by each organ and calculated using the RODES software [3, 4].

References

[1]. International Commission on Radiological Protection.. Oxford: Pergamon Press; ICRP Publication 94 (2004).

---

*Corresponding author: m_mezaguer@yahoo.fr

[2]. United Nations Scientific Committee on the Effects of Atomic Radiation. Sources and effects of ionizing radiation. 2000 report to the general assembly with annexes. Vienna: United Nations; 2001.

[3]. M. Mezaguer-Lekouaghet, A. Badreddine, N. Ait Kaci, K. Remil, A. Benbournane, F. Boucheneb, F. Messili, A. Baz, M. Souidi, & Z. Lounis-Mokrani, ICNTRM. Strasbourg France (2017).

[4]. M. Mezaguer-Lekouaghet, Eric Blanchardon, A. Badreddine, Jean-Marc Bertho, Maamar Souidi, Ahcène Baz & Z. Lounis-Mokrani. (It is presented in the same conference).

BIO Web of Conferences **14**, 05010 (2019) https://doi.org/10.1051/bioconf/20191405010
HEIR 2018

Toxicity mechanisms of cobalt oxide particles (Co_3O_4P) on human lung cells: impact of solubilization

Véronique Malard[1,*], *Carole* Bresson[2], *Chiara* Uboldi[3], *Pascale* Delangle[4], *Anne* Van der Meeren[5], *David* Lemaire[1], *Carine* Darolles[6], *Thierry* Orsière[3], *Richard* Ortega[7], and *Catherine* Berthomieu[1]

[1]Laboratory of Protein-Metal Interactions (LIPM), Institute of Biosciences and Biotechnologies of Aix Marseille (BIAM), UMR7265 CEA – CNRS - Aix Marseille Univ, CEA Cadarache, F-13108 Saint-Paul-lez-Durance, France

[2]Den – Service d'Etudes Analytiques et de Réactivité des Surfaces (SEARS), CEA, Université Paris-Saclay, F-91191, Gif sur Yvette, France

[3] Institut Méditerranéen de Biodiversité et d'Ecologie marine et continentale (IMBE), Aix Marseille Univ, CNRS, IRD, Univ Avignon, Equipe Biomarqueurs, Environnement Santé, Faculté de Médecine Timone, Marseille, France

[4] INAC-SyMMES, Univ. Grenoble Alpes, CEA, CNRS, 38000 Grenoble, France

[5] Laboratoire de Radiotoxicologie, CEA, Paris-Saclay University, 91297 Arpajon – France

[6] CEA, DRF, Biosciences and biotechnologies institute (BIAM), Bagnols-sur-Cèze, France

[7]Centre d'Etudes Nucléaires de Bordeaux Gradignan (CENBG), CNRS, IN2P3, Université de Bordeaux, UMR 5797, F-33170 Gradignan, France

[*] Corresponding author: veronique.malard@cea.fr

The wide use of cobalt in industrial applications, including the nuclear industry, could lead to accidental or chronic occupational exposure mainly by inhalation. Cobalt may be encountered in industry in a variety of chemical forms, including metal dusts, oxides (Co_3O_4), and soluble salts and could also be present in fragments of irradiated fuel. ^{60}Co is an important activation product produced in nuclear power plants. The potential toxicity of poorly soluble cobalt (II, III) oxide particles (Co_3O_4P) is a consequence of the long retention time (years) in lungs of the gamma-ray emitting radionuclide. Lung damage may be expected from both chronical radiation exposure, and chemical toxicity of Co.
As a first approach to evaluate the chemical toxicity of cobalt particles, stable $^{59}Co_3O_4$ particles were used. We have demonstrated that cobalt oxide particles were readily endocytosed by the BEAS-2B human lung cells via the clathrin-dependent pathway. Despite their very low solubility in the culture medium, cobalt particles were partially solubilized at low pH within lysosomes, leading to the release of cobalt ions. Using ion beam microanalysis, solubilized cobalt was detected within the cytoplasm and the nucleus. As expected with these low-solubility particles, the intracellular solubilized cobalt content was low compared with the intracellular particulate cobalt content in the parts-per-thousand range or below. However, this minute fraction of intracellular solubilized cobalt was found to be responsible for the overall toxicity [1].
The genotoxic potential of poorly soluble Co_3O_4P was then investigated in the BEAS-2B cell line, which exhibits the highest homology in gene expression pattern with primary non

tumor lung cells. Cytokinesis-block micronucleus assay and comet assay revealed that Co_3O_4P had a genotoxic potential following exposure at non cytotoxic concentrations. Indeed, enhanced micronuclei formation and both primary and oxidative DNA damage were noted following exposure to increasing concentrations of Co_3O_4P. The scoring of γ-H2Ax foci demonstrated that Co_3O_4P were able to generate double DNA strand breaks. The involvement of oxidative stress was confirmed by the inhibition of foci formation in the presence of the ROS scavenger N-acetylcysteine. Comparison of the effects induced by Co_3O_4P with those triggered by Co^{2+}, indicated that the genotoxicity of Co_3O_4P was not solely due to the released soluble fraction [2].
To go further, a project aiming at broadening the therapeutic offer in case of accidental contamination with insoluble cobalt particles was recently undertaken. This project, dealing with $^{60}Co_3O_4$, uses a transversal approach ranging from chemistry (to select ligands able to enhance particle solubilization), to decorporation studies in rodents. Preliminary results obtained in the framework of this collaborative project will be presented.

References

[1] Ortega R, Bresson C, Darolles C, Gautier C, Roudeau S, Perrin L, Janin M, Floriani M, Aloin V, Carmona A, Malard V. *Low-solubility particles and a Trojan-horse type mechanism of toxicity: the case of cobalt oxide on human lung cells*. Part Fibre Toxicol. 2014; 11 (1) 14.
[2] Uboldi, C., Orsiere, T., Darolles, C., Aloin, V., Tassistro, V., George, I., & Malard, V. *Poorly soluble cobalt oxide particles trigger genotoxicity via multiple pathways*. Part Fibre Toxicol. 2016, 13 (1), 5.

Results presented here were obtained in the frame of the EDF-CEA collaboration.

BIO Web of Conferences **14**, 05011 (2019) https://doi.org/10.1051/bioconf/20191405011
HEIR 2018

Differential responses to tritiated thymidine incorporation into DNA between mouse embryonic neural stem cells and fibroblasts

Sofiane Mokrani[1], *Christine* Granotier-Beckers[1,*], *Olivier* Etienne[1], *Christian* Grisolia[2], and *François* Boussin[1]

[1]CEA, JACOB, IRCM, INSERM U967, Laboratoire de Radiopathologie, Université Paris Diderot Sorbonne Paris Cité, Université Paris XI, 92260 Fontenay-aux-Roses, France
[2]CEA, DRF, IRFM, SI2P, GCFPM, 13115 Saint-Paul-lez-Durance, France

Tritium is the radioactive isotope of hydrogen with a low beta energy emission. It is generated naturally or released by nuclear power plants. Tritium has a persistent and ubiquitous distribution in the environment and different biological systems.

Epidemiological data and experiments on animal models have shown the damaging effects of tritiated compounds [1-3], but their impact on the developing brain remains poorly known. The developing brain is particularly radiosensitive because of the genetic damage induced in the Neural Stem and Progenitor Cells (NSPCs) which generate most of brain cells.

Previous results in the laboratory have described the cellular and genetic effects of tritiated thymidine (^{3}H-T) on primary cultures of human embryonic NSPCs. A better understanding of the molecular mechanisms involved in the genotoxicity of tritium is now required to define the individual sensitivity factors to tritium incorporation.

In this study, we used primary cultures of NSPCs isolated from mouse embryonic brain cortex after 14 days of gestation. We characterized the chromosome instability induced by increasing doses of ^{3}H-T incorporation into the DNA during replication. Our results show that mouse and human embryonic NSPCs exhibit similar responses to ^{3}H-T incorporation.

Primary cultures of Mouse Embryonic Fibroblasts (MEF) were also isolated from skin at the same developmental stage.

Using cellular autoradiography we revealed that ^{3}H-T induces high chromosome instability in MEF than NSPCs. We also observed significantly higher radiosensitivity of MEF as compared to NSPCs after γ-radiation exposure.

These data evidence significant differences in the DNA damage responses of MEF and NSPCs. In order to further investigate the mechanisms involved, we have compared the chromosome instability induced by tritium incorporation in NSPC and MEF obtained from mice deficient for different DNA repair pathways.

References

* Corresponding author: christine.granotier@cea.fr

1. S. Zamenhof, E. van Marthens, Radiat Res, **77** (1979)
2. SJ. Bursian, DF. Cahill, JW. Laskey, LN. Parker, Int J Radiat Biol Relat Stud Phys Chem Med, **27**,5 (1975)
3. B. Wang, H. Takeda, WM. Gao, XY. Zhou, T. Odaka, H. Ohyama, Health Phys. **77**, 1(1999).

Differential effects of HTO and OBT ingestion or external irradiation on iron metabolism

Jean-Marc Bertho[1,*], *Dimitri* Kereselidze[1], *Line* Manens[1], *Cécile* Culeux[1], *Joel* Surette[2], *Melinda* Blimke[2], *Linsday* Bertrand[2], *Heather* Wyatt[2], *Maamar* Souidi[1], *Nick* Priest[2], *Jean-René* Jourdain[1].

[1] IRSN, PSE-SAN/SESANE, LRTOX, BP 17, 92262 Fontenay aux roses cedex, France

[2] Canadian Nuclear Laboratories Ltd, Chalk River, ON, Canada.

1 Introduction

Environmental exposure to tritium is mainly due to tritium releases by nuclear reactors. This exposure may increase in future years due to potential increased releases by the ITER reactor. However concerns exist about the reliability of the current ICRP recommendations. In fact, the current regulatory level is set at 7000 $Bq.L^{-1}$ in Canada, WHO recommendation is set at 10 $kBq.L^{-1}$ and French regulatory level is set at 100 $Bq.L^{-1}$. As a result, a collaborative project was developed between IRSN and CNL to study various aspects of the toxicity of tritium.

2 Material and methods

In this project male C57BL/6J mice were exposed to a chronic contamination through drinking water with three different concentrations of tritium (HTO), or tritiated amino acids (TAA) or low dose external gamma radiation (1.3 and 31.2 μGy.h-1), equivalent to tritium exposure, for either 1 or 8 months. Drinking water concentrations of tritium were 10 $kBq.L^{-1}$, 1 $MBq.L^{-1}$ and 20 $MBq.L^{-1}$. The lowest concentration used in this study is relevant to WHO recommendations. At 1 and 8 months of exposure, blood, spleen, liver, kidney and intestine were harvested and subjected to various analyses including blood cell count and differential, measurement of biochemical parameters in the plasma, relevant genes expression quantification and relevant immunohistological analyses.

3 Results

Results of blood cell count analyses indicated slight but significant decrease in RBC at 1 month of exposure with TAA which was followed at 8 months by an increase in the size of RBC. These changes were not observed with HTO exposure. These results suggest the appearance of an anaemia at 1 month of exposure with TAA, which was compensated by an increased half-life of red blood cells at 8 months of exposure. A concomitant decrease in serum iron concentration at 8 months of exposure in TAA-exposed animals was observed

* Corresponding author: jean-marc.bertho@irsn.fr

with a dose dependent effect suggesting that an iron deficit was at the origin of this anaemia. In fact, the increased life span of red blood cells is a classically described compensation mechanism for a mild iron deficit. Such an iron-dependent anaemia may be due to either a central defect of haematopoiesis, a defect in EPO production by the kidneys, a defect in the capture of RBC in the spleen or a defect in iron metabolism. Results of gene expression analysis, histological analysis and cytokine measurement allowed excluding the three first hypotheses. However, a decrease in ERFE expression in the liver at one month and a decrease in transferrin light chain expression in intestine, both at 1 month and 8 months of exposure were evidenced. ERFE is implicated in a regulation loop of iron concentration in the blood through the inhibition of Hepcidin, and transferrin is implicated into the intracellular iron transport. Interestingly, a change in the pattern of histological localization of DMT1 was observed in the intestinal villi, suggesting that the iron transport through the intestinal epithelium is the key parameter that may explain these results. Work is currently in progress to confirm these results.

4 Conclusions

Overall these results demonstrate a differential effect of HTO and OBT at the same level of exposure. In parallel, no effect was observed in animals gamma irradiated at a low dose rate equivalent to the dose rate due to incorporated tritium. This suggests that deterministic effects may appear during chronic ingestion of tritium and that these effects may be significantly more important with an OBT form of tritium as compared to HTO. This may be explained by the integration of TAA in the general metabolism, and more specifically in the metabolism of intestinal epithelial cells. These results also suggest that the effect of TAA might be dependent upon the physiology of the amino acids used in this study, i.e. on the specific pattern of exposure to tritium beta rays according to the micro anatomic location of these amino acids. These results also suggest that the biological effects of other forms of OBT, e.g. tritiated sugars or tritiated lipids, might be very different. The implication of these results for radioprotection remains to be defined.

BIO Web of Conferences **14**, 05013 (2019) https://doi.org/10.1051/bioconf/20191405013
HEIR 2018

Effects of tritiated thymidine on human neural stem and progenitor cells

Christine Granotier-Beckers[1,*], *Sofiane* Mokrani[1], and François D. Boussin[1]

[1]CEA, JACOB, IRCM, INSERM U967, Laboratoire de Radiopathologie, Université Paris Diderot Sorbonne Paris Cité, Université Paris XI, 92260 Fontenay-aux-Roses, France

Ionizing radiation during pregnancy may impair brain development resulting in microcephaly and mental retardation. These effects might be related to the targeting of highly proliferating NSPC during brain development. These cells give rise to the main types of neural differentiated cells of the brain.

Tritium which is the radioactive hydrogen isotope emitting β-particles is either of natural origin or is produced by nuclear facilities for fusion and fission reactions. Tritium dispersion around nuclear facilities may result in people exposure to very low doses of tritium in different chemical forms (tritium gas, tritiated water or organically bound tritium).

Today, low dose exposition to tritium and its consequences on human health are still poorly known, in particular for the developing brain. By contrast, previous studies have shown that prenatal exposition to high doses of tritiated water or tritiated thymidine induce microcephaly and neurobehavioral changes in rat and mice models [1-3].

We have thus investigated the effects of tritiated thymidine on primary cultures of human neural stem and progenitor cells from several donors. Tritiated thymidine is incorporated in DNA during replication leading to chronic irradiation of the contaminated cells and their progeny.

We have shown that tritiated thymidine incorporation led to dose-dependent decrease of cell proliferation in NSPC. This effect was due to inhibition of cell cycle progression and induction of apoptosis and premature senescence. Tritiated thymidine also impaired neural stem cells functions by triggering their premature differentiation.

Intracellular tritium dosimetry based on autoradiography revealed that cellular effects were only associated to high levels of tritiated thymidine incorporation.

We then compared the chromosome instability induced by tritiated thymidine after several rounds of replication. Cytogenetic analysis of metaphase spreads evidenced that tritiated thymidine induced chromatid-type aberrations after the first round of incorporation, consistently with the tritiated thymidine incorporation in the newly synthesized DNA strands. Tritium dosimetry revealed that one chromatid break was detected for about 50 tritium disintegrations occurring during S and G2/M phases.

Both chromatid-type and chromosome-type aberrations are detected in γ-irradiated cycling cells, resulting from DNA damage induced before or after DNA replication respectively. Interestingly, chromatid-type aberrations were still mainly induced by tritiated

* Corresponding author: christine.granotier@cea.fr

thymidine incorporation during several cell cycles whereas very few chromosome-type aberrations were observed. Moreover, the number of chromosome aberrations did not increase along the successive rounds of tritium incorporation, suggesting that contaminated cells developed an adaptive response. Further studies are required to investigate the molecular mechanisms involved.

References

1. S. Zamenhof, E. van Marthens, Radiat Res, **77** (1979)
2. SJ. Bursian, DF. Cahill, JW. Laskey, LN. Parker, Int J Radiat Biol Relat Stud Phys Chem Med, **27**,5 (1975)
3. B. Wang, H. Takeda, WM. Gao, XY. Zhou, T. Odaka, H. Ohyama, Health Phys. **77**, 1(1999).

Exploiting the " Rat Actinide Repository " at the Laboratory of RadioToxicology, CEA, France

Nina M. Griffiths[1*], *Stéphanie* Lamart[1], *Anne-Claire* Humbert[2], *Anne* Van der Meeren[1]

[1]Laboratoire de RadioToxicologie, CEA, Université Paris-Saclay, 91297 Arpajon, France

[2] CEA, DAM, DIF, 91297 Arpajon, France

1 Introduction

The instigation of "registries" and "repositories" of biological samples from workers in the nuclear industry, both civil and military, provides an excellent source of data concerning the long-term effects of internal contamination by radionuclides (e;g. U.S. Transuranium and Uranium Registries, USTUR; Radiobiological Human Tissue Repository, RHTR). Similarly tissue archives exist concerning animal experiments. During many years of research dating from the 1960s at the Laboratory of RadioToxicology (LRT) a large number of tissue samples were collected from animals contaminated with alpha particle-emitting actinides, mostly following inhalation. To date around 3000 samples and 7000 unused sections (5 µm) from 1500 rats have been included in the inventory of samples at LRT. Samples are from a number of studies such as (i) pulmonary inflammatory responses and pathologies following inhalation of MOX [Mixed OXide, Uranium, plutonium (Pu)] oxide or Pu oxide, (ii) osteosarcomas following injection of soluble Pu and biokinetics and (iii) effects of different regimens for decorporation following contamination of wounds.

2 Exploitation of samples from the repository: examples

The formalin-fixed paraffin-embedded (FFPE) tissues have been mainly used for histopathological analyses and in particular for description of lung tumour types following inhalation of actinides (MOX, plutonium, neptunium). In addition inflammatory responses and fibrosis were studied in lungs from MOX-contaminated animals[1]. However these tissues may be used for other analytical purposes – such as identification of different elements or isotopes in particular tissue structures. For example, using lungs obtained from MOX-contaminated rats (Fig. 1), fibrotic areas containing aggregates of activity were identified by emulsion and solid track (CR39) autoradiography and dissected from sections (60 µm thick) and subjected to further analyses. The alpha spectrophotometry measurements show that the aggregates formed in the fibrotic tissue contain both Pu and Am and importantly, in the same expected isotopic/elemental ratios as calculated from isotopic decay of the particular MOX powder. Of note is the "in-growth" of Am. In animal experiments the time of contamination is known, the contaminant is known and so calculations of Am "in-growth" can be fairly exact. In the analysis of tissues from cases of accidental actinide contamination this may not necessarily be the case. It is important

* Corresponding author: nina.griffiths@cea.fr

therefore to consider this when tissues are analysed many years after the actual contamination.

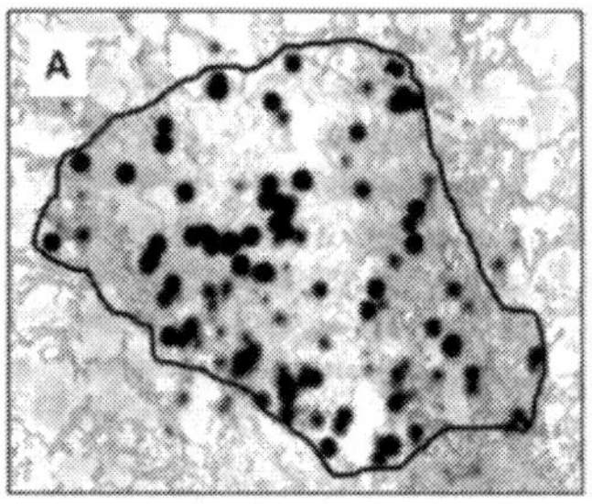

B Analysis of tissue Pu and Am

$^{239}Pu + ^{240}Pu$	^{238}Pu	^{241}Am	Total
0.7 Bq	1.5 Bq	0.8 Bq	3.0 Bq
23 %	50 %	27 %	(Measured)
MOX Pu – Am at time of inhalation experiment (2004)			
19.3 %	*61.9 %*	*18.7 %*	*(Calculated)*
MOX Pu – Am at time of tissue analysis (2010)			
17.8 %	*54.5 %*	*27.6 %*	*(Calculated)*

Fig. 1. A: Autoradiograph (emulsion) of rat lung section with aggregates of activity in a fibrotic zone after inhalation of MOX (7,1 % Pu). Counter staining haematoxylin-eosin. The delimited area was dissected using a microscope and fine needles. The tissue was weighed and analysed for Pu and Am following column separation and alpha spectrometry. **B:** Table showing measured Pu or Am activity in the dissected tissue and the different *calculated isotopic fractions* of MOX either at the time of inhalation (2004) or time of tissue analysis (2010).

More recent experiments using the iQID (ionizing-radiation Quantum Imaging Detector) camera where activity can be measured directly on a tissue section results showed that in lungs from MOX-contaminated rats, higher levels of activity were found in areas of fibrosis as compared with non-fibrotic regions[2]. This technique however has the advantage of being non-destructive and measures activity directly as opposed to dissection and lengthy destructive radiochemical procedures. To date however, differential isotopic analysis is not possible. Distinct lung pathologies (fibrosis, tumours) are consequential to inhalation of mainly insoluble actinides but following systemic administration of more soluble Pu forms osteosarcomas are the main result. Osteosarcoma FFPE samples from the LRT Rat Actinide Registry/Repository were analysed for different tumour-associated markers. The constitutive activation of the Wnt/beta-catenin pathway in these tumours was identified by nuclear and/or cytoplasmic localization of beta-catenin and a decrease in its inactive (phosphorylated) form in tissue sections[3].

3 Conclusions

These tissues obtained from different contamination protocols with different actinides provide a source of material that may be exploited in a number of ways – histopathological analyses with new markers for example of proliferative lesions such as cancer-associated fibroblasts, activity measurements at a microscopic level by SIMS, comparisons of Am "in-growth, nuclear forensic analyses of plutonium isotopes… . This tissue repository is a legacy for radiotoxicological measurements and also a significant resource for future that fulfil the remit of the 3Rs (Replace, Reduce, Refine) for animal experimentation.

The authors are very grateful to Valérie Quéméneur for excellent organisational and cataloguing skills for the recent inventory of the tissue blocks and to the EU for financial support (MELODI).

References

1. N.M. Griffiths, A. Van der Meeren, P. Fritsch, M-C. Abram, J-F. Bernaudin, J.L. Poncy, Health Phys. **99**(3): p. 347-56 (2010).
2. S. Lamart, B.W. Miller, A. Van der Meeren, A. Tazrart, J.F. Angulo, N.M. Griffiths, PLoS One. **12**(10):e0186370 (2017).
3. K. Daino, N. Ugolin, S. Altmeyer-Morel, M-N. Guilly, S. Chevillard, Int J Cancer. **125**(3): p. 612-20 (2009).

Distribution of uranium in the internal organs of laboratory animals after incorporation of uranium ore dust

Yelena Saifulina[1], *Sevda* Toklayeva[1], *Maxut* Sharipov[1], *Riza* Medethan[1], *Meirat* Bahtin[1], and *Polat* Kazymbet[1]

[1]JSC Astana medical university/ institute of radiobiology and radiation protection

The aim of the study was to assess the levels of uranium concentration in tissues and organs of laboratory animals by oral ingestion of uranium ore dust into the body.

A series of studies was performed on 43 mature white mongrel rats of both sexes. The animals were divided into two groups: the main group (n = 30), where laboratory rats were mixed with uranium ore dust (UOD) at a dose of 50 maximum allowable a concentrations (MAC). Group II (n = 13) comprised intact animals. The oral intake of the(UOD) was simulated for 7 days of daily feeding of white outbred rats by the dust of uranium ore mixed with the feed. The dust mass of uranium ore for single feeding of animals of the experimental group at a dose of 50 MPC was 5750 mg. Biomaterials were taken after end of oral irradiation 24 hours after the last feeding of UOD. The concentration of uranium in the organs and tissues of rats was measured on an inductively coupled plasma mass spectrometer "Agilent-7800" with a range of measured masses from 2 amu. up to 260 amu Measurements of the specific activity of ^{226}Ra, ^{232}Th in rat feces were performed daily after each feeding by spectrometry on a device using the gamma-beta spectrometer Progress-BG.

The content of uranium was measured in the organs of the digestive system: the small intestine and the liver. The average concentration of uranium in the small intestine was 2306.2336.6 ng / l and exceeded the corresponding values in the control group by more than 50 times. In the liver, this indicator was also high and 8 times different from the amount of uranium in the control animals - 451.9-315.9 ng / l. The average uranium content in the kidneys of irradiated rats was 18623.8-4326.4 ng / l, which significantly differed from the concentration of uranium in the kidneys of intact rats by more than 500 times. In the brain of irradiated rats, the uranium concentration was almost 8 times higher than in the control group and amounted to 17.14 ng / l.

The average value of the specific activity of ^{226}Ra released in feces during the period of admission of UOD into the body of rats was 912.2-460.0 Bq / kg, with its greatest activity after 2 days of feeding (1420.5-154.9 Bq / kg). In the subsequent week after the termination of internal irradiation, the activity of radium decreased to 211.9±106.4, but did not reach the activity values in the control group (8.3 Bq / kg). A decrease in the specific activity of ^{232}Th in animal feces did not occur higher values were recorded the following week after feeding (624.9±223.1 Bq / kg) than during the feeding period of OUD (493.8±192.9 Bq / kg).

The results of the experiment demonstrate a faster rate of removal of from^{226}Rathe body of animals through the gastrointestinal tract than^{232}Th. In the early periods after internal oral irradiation of UOD animals, the maximum concentrations of uranium are determined in the kidneys, small intestine and liver.

Medical countermeasures against radionuclide contamination: An overview

Olivier Grémy*

Laboratoire de Radiotoxicologie, CEA, Université de Paris-Saclay, Bruyères-le-Châtel, France

1 Introduction

Radionuclides (RN) are extensively used around the world in a broad spectrum of industrial (e.g. electric power generation), medical (e.g. therapies, imaging), and military applications (e.g. armor-piercing weapons), thus causing opportunities for accidental intake by individuals. Although this hazard primarily concerns workers and soldiers, the population can also be affected in the case of major nuclear accidents resulting in a large release of RN into the environment, or following the possible terrorist use of a RN dispersal device.

Internal contamination of RN can result from inhalation, ingestion, wound or penetration through intact skin. Following internalization, a fraction of RN will be absorbed into the blood prior to deposition in target organs. During their retention which depends on their biological half-lives, RN have the potential to cause radiation-induced health effects or to even impair physiological systems.

RN of health concern are actinides, especially uranium (U) and plutonium (Pu) isotopes, but also americium (^{241}Am), neptunium (^{237}Np) and curium isotopes (Cm), the fission and activation products iodine-131 (^{131}I), cesium-137 (^{137}Cs), strontium-90 (^{90}Sr), tritium (^{3}H), and cobalt-60 (^{60}Co).

2 Current treatments and recommendations

The only practical way to reduce RN body burden and associated health risks is to apply countermeasures, i.e. decorporation and decontamination methods, that consist in minimizing blood absorption and/or accelerating excretion of RN. The therapeutic arsenal includes non-specific methods. For ingested RN, intestinal absorption can be reduced by gastric lavage, enemas or by gastric dressing with antacids. An intact skin or wound can be decontaminated by washing with tepid water, a decontaminant solution or soap (e.g. Trait rouge®), or applying Osmogel®, green clay... A surgical excision of the wound site will be considered for a puncture wound with high specific activity RN. In case of massive inhalation of insoluble forms of highly radioactive transuranics such as ^{238}Pu oxides, the removal of macrophage-internalized RN particles by lung lavages could be envisaged.

Other strategies are more specific. The countermeasure against ^{131}I is the oral intake of potassium iodide tablets to prevent thyroid uptake by saturating binding sites. Similarly, the method of isotopic dilution with a stable isotope or a chemical analog is recommended for competing with the RN, so accelerating excretion. Sr lactate or calcium gluconate can be

* Corresponding author: olivier.gremy@cea.fr

given parenterally for ^{90}Sr contamination, and drinking large amounts of water can facilitate tritium clearance. Also, Co oligosol administration may be useful to accelerate ^{60}Co body turnover. For U intake, the administration of sodium bicarbonate *per os* or by i.v. infusion is recommended. This can promote the formation of U-carbonate complexes in blood that are stabilized by urine alkalization, thereby preventing kidney deposition of U [1]. The medication against ^{137}Cs ingestion is the long term oral treatment with insoluble Prussian Blue (Radiogardase®) that contains crystals of ferric ferrocyanide. This traps available Cs in gut by ion-exchange so minimizing intestinal absorption and reabsorption by blocking Cs enterohepatic circulation. The only available decorporation therapy for transuranics Pu, Am or Cm is the i.v. infusion of the marketed solution containing the octadentate chelator diethylenetriaminepentaacetic acid (DTPA) as calcium or zinc trisodium salts. Inhalation of nebulized DTPA is authorized for treating inhaled RN and irrigation with the solution is recommended for skin or wound decontamination. EDTA or DTPA are sometimes recommended for Co since some animal studies have shown a modest efficacy.

3 Animal research for improvement of decorporation treatment

There is a vast body of studies which aims at improving treatments mainly with chelators, either by finding or developing new ones, or by reformulating those available to enhance systemic efficacy, to target the site of RN entry, or to facilitate administration.

3.1 Evaluation of already-marketed drugs or available natural compounds

Natural compounds or licensed drugs have been evaluated in animals for decorporating RN, such as the amino acid N-AcetylCysteine (NAC), the tripeptide L-glutathione (GSH), a dithiol lead chelator (dimercaptosuccinic acid, DMSA), copper chelators (D-penicillamine and trientine) and polysaccharides. In many cases, results are mixed or modest. A few examples. Injections of GSH or NAC may decrease hepatic and splenic contents of Co [2]. Of interest an older study showed that a repeated treatment with DTPA was better when the chelator given as Co salts rather than as Ca or Zn salts [3]. Chitosan lactate, cysteine and D-penicillamine may also have potential for Co decorporation [4-6]. Alginate can remove ingested Sr [4]. However, phytate (Zn-InsP6) seems to inhibit more effectively Sr intestinal absorption [7]. Trientine and D-penicillamine modestly increase the clearance of Sr and Cs, respectively [6]. The consumption of pectins to eliminate the chronically ingested Cs may be useful as shown by some studies conducted on contaminated children [8, 9]. As for Pu, chelators primarily designed for treating iron overload such as desferrioxamine have been tested but none is better than DTPA. An active substance of drugs for treating Ca-related bone diseases, the bisphosphonate etidronate (EHBP), limits U retention [10, 11].

3.2 Approaches for high affinity chelating agents

For Pu decorporation, derivatives of DTPA have been designed by altering their structure, adding or replacing chelating moieties, e.g., the dihydroxamic derivative DTPA-DX. Nonetheless none have proved to be more effective than native DTPA.

Considering the similar chemical properties and the close *in vivo* behavior of Fe(III) and Pu(IV)), numerous chelators have been synthetized by varying the molecular backbones and geometry, as well as the siderophore-inspirated chelating moieties and number (denticity) [12]. Based on their decorporation efficiency and their toxicity in animals [13, 14], two of those synthetic chelators have emerged as lead candidates due to their far greater efficacy than DTPA for Pu removal and their ability to sequester other actinides including Am, Np and U: the octadentate 3,4,3-LI(1,2-HOPO) and the tetradentate 5-LIO(Me-3,2-HOPO), both containing hydroxypyridinone chelating moieties grafted to linear polyamine backbones. More recently, 3,4,3-LI(1,2-HOPO) was preferred to take forward through pre-clinical studies of efficacy, toxicology, and pharmacology [15]. Some chelators bearing catecholamide functions such as 3,4,3-LICAM(C) for Pu removal, or 5-LICAM(S) and CBMIDA for U [10, 14] were also promising but were later shown to be nephrotoxic and less efficient than HOPO-based chelators.

The affinity of a phosphonic acid moiety for U has been also an area of research for U decorporation improvement. For example, dipodal and tripodal U-binding diphosphonate chelators with amine as scaffold have been synthetized [16]. The most potent for reducing U retention was the dipodal designated 3C which is slightly better than EHBP but less efficient than 5-LICAM(S) for bone deposits.

Chelators with macrocyclic structures have been also tested. Thus, for decontamination of intact skin, a calix[6]arene bearing carboxylic groups (p-tert-butylcalix[6]arene) has been formulated in an oil-in-water emulsion [17]. It is efficient for U removal but may have the ability to trap other actinides. Heterocyclic crown ether compounds can promote body clearance and reduce tissue retention of Sr [18].

The use of functionalized macromolecules bearing numerous chelating sites per area unit is under investigation. For example, the functionalized methy-carboxylated poly(ethylenimine) (PEI-MC) has an high affinity for U but no animal studies have been performed yet [19]. Also, a mesoporous silica (SAMMS^{TM}) with ferrocyanide moieties has a greater capacity for Cs loading than Prussian Blue but is not more efficient *in vivo* [20].

3.3 Strategies for oral delivery enhancement of chelators

The parenteral route for treating many people would be challenging and make a long-term therapy particularly cumbersome for patients. Orally effective chelators would be very useful but the available DTPA is very poorly absorbed from the intestine (3-5%).

Partially lipophilic analogs and derivatives of DTPA were developed to facilitate passage through the intestinal barrier. Long alkane chains were grafted to a nonadentate analog of DTPA named triethylenetetramine hexaacetic acid (TTHA), resulting in a series of compounds of which the most interesting is $C_{22}TT$ for Pu/Am decorporation [21]. Another approach was to create a prodrug such as the penta-ethyl ester (C2E5) of DTPA [22]. In addition, pharmacological approaches have been employed to enhance intestinal penetration of native DTPA. Indeed, DTPA associated with permeation enhancers as a tablet form [23] or entrapped within enteric-coated capsules [24] have been formulated. Also, a water-in-oil microemulsion of DTPA has been developed for buccal transmucosal delivery (Medesis Pharma; personal data). However, all these DTPA galenic forms need repeated administrations at higher doses to be as effective as parenteral DTPA. We note that the aforementioned HOPO-based chelators already display a good oral efficiency [13] which could be undoubtedly improved further by a permeation enhancer.

3.4 Strategies targeting the site of entry or a systemic retention tissue of RN

A rapid delivery of chelators at the site of entry (lung, wound site and intestine) aims at preventing RN absorption into the blood as non-chelated form, which will lead to a better overall decorporation than a systemic delivery that must trap RN in circulation or in tissue.

Thus, DTPA has been formulated into porous particles by spray-drying for improving local treatment in case of inhaled RN. Its insufflation in rats induced a good mobilization of available Pu deposited in lungs and prevented deposition in systemic tissues [25].

Along with bone, the liver is a target organ for transuranics. The encapsulation of DTPA within liposomes allows better penetration into hepatic cells so improving decorporation for liver-retained Pu [26].

4 Conclusion

The treatments for RN contamination are restricted to a small number of non-specific methods and strategies using specific chelators. Some RN remain without any or really satisfactory treatment such is the case for Np, Co and polonium.

So far, work has focused mainly on Pu and U, while considerably less study has been directed towards other RN. Efforts still need to be made to complete the poor therapeutic arsenal against RN internalization with efficient drugs. Today, the best candidate is undoubtedly 3,4,3-LI(1,2-HOPO) with a greater efficacy of Pu/Am decorporation than DTPA, as well as an ability to chelate U and Np. It received approval by the US FDA for phase I clinical trials in 2014.

Pending approval and licensing of new chelators, the development of novel drug-delivery systems for already available chelators to increase their bioavailibility may be a strategy for decorporation improvement, and simpler for attainment of regulatory approval.

References

1. Y. Ohmachi, T. Imamura, M. Ikeda, E. Shishikura *et al.*, J Toxicol Pathol, **28** (2015)
2. J. M. Llobet, J. L. Domingo and J. Corbella, Arch Toxicol, **58** (1986)
3. D. R. Fisher and B. G. Dunavant, Health Phys, **35** (1978)
4. T. G. Levitskaia, J. A. Creim, T. L. Curry, T. Luders *et al.*, Health Phys, **99** (2010)
5. T. G. Levitskaia, J. A. Creim, T. L. Curry, T. Luders *et al.*, Health Phys, **97** (2009)
6. T. G. Levitskaia, J. A. Creim, T. L. Curry, T. Luders *et al.*, Health Phys, **101** (2011)
7. K. Ogawa, M. Aoki, S. Kadono and A. Odani, PLoS One, **13** (2018)
8. V. B. Nesterenko, A. V. Nesterenko, V. I. Babenko, T. V. Yerkovich *et al.*, Swiss Med Wkly, **134** (2004)
9. P. Hill, M. Schlager, V. Vogel, R. Hille *et al.*, Radiat Prot Dosimetry, **125** (2007)
10. S. Fukuda, H. Iida, M. Ikeda, X. Yan *et al.*, Health Phys, **89** (2005)
11. M. H. Henge-Napoli, E. Ansoborlo, V. Chazel, P. Houpert *et al.*, Int J Radiat Biol, **75** (1999)
12. A. E. Gorden, J. Xu, K. N. Raymond and P. Durbin, Chem Rev, **103** (2003)
13. R. J. Abergel, P. W. Durbin, B. Kullgren, S. N. Ebbe *et al.*, Health Phys, **99** (2010)
14. P. W. Durbin, Health Phys, **95** (2008)
15. T. A. Choi, A. M. Furimsky, R. Swezey, D. I. Bunin *et al.*, J Pharm Sci, **104** (2015)
16. M. Sawicki, D. Lecercle, G. Grillon, B. Le Gall *et al.*, Eur J Med Chem, **43** (2008)
17. G. Phan, N. Semili, C. Bouvier-Capely, G. Landon *et al.*, Health Phys, **105** (2013)
18. T. Colomina, J. M. Llobet, J. L. Domingo and J. Corbella, Vet Hum Toxicol, **33** (1991)
19. F. Lahrouch, O. Sofronov, G. Creff, A. Rossberg *et al.*, Dalton Trans, **46** (2017)
20. C. Timchalk, J. A. Creim, V. Sukwarotwat, R. Wiacek *et al.*, Health Phys, **99** (2010)
21. S. C. Miller, X. Wang and B. M. Bowman, Health Phys, **99** (2010)
22. K. Sueda, M. P. Sadgrove, J. E. Huckle, M. G. Leed *et al.*, J Pharm Sci, **103** (2014)
23. G. N. Shankar, W. Weber, M. Doyle-Eisele and R. A. Guilmette, Drug Dev Res, **73** (2012)
24. J. P. Wilson, R. R. Cobb, N. W. Dungan, L. L. Matthews *et al.*, Health Phys, **108** (2015)
25. O. Gremy, N. Tsapis, S. Bruel, D. Renault *et al.*, Radiat Res, **178** (2012)
26. O. Gremy, L. Miccoli, F. Lelan, S. Bohand *et al.*, Radiat Res, **189** (2018)

Liposomal DTPA as a good strategy for enhancing Pu decorporation regardless of treatment regimen.

Olivier Grémy[1,*], *Laurent* Miccoli[1], *Faustine* Lelan[1], *Sandra* Bohand[2], and *Marie* Mougin-Degraef[3]

[1]Laboratoire de Radiotoxicologie, CEA, Université de Paris-Saclay, Bruyères-le-Châtel, France
[2]Orano Mining, Courbevoie, France
[3]CRCINA, Inserm, CNRS, Université d'Angers, Université de Nantes, Nantes, France

1 Introduction

Internal contamination with plutonium (Pu), either in the context of accidental occupational exposure, after a natural disaster affecting nuclear facilities, or as a result of possible terrorist use, continues nowadays to be a potential hazard. The only practical way to reduce Pu body burden and the associated radiation risks is decorporation by chelation therapy, which consists in administering a saline solution of the chelating agent diethylenetriaminepentaacetic acid (DTPA used as Ca or Zn trisodium salt). Unfortunately, the very short biological half-life (90-99% excreted within 24 hours) and the too low ability to enter cells of DTPA seriously limit its efficacy for Pu decorporation.

The encapsulation of DTPA within liposomes, i.e., artificial lipid spherules, provides a greater cell penetration and a longer retention of DTPA in tissues, thereby enhancing its ability to mobilize various deposited metals including Pu, compared to free DTPA. Most of the decorporation studies have assessed the efficacy of liposomal DTPA on animals injected with polymers/colloids of Pu.

The present study aimed at assessing the efficacy of 110nm-sized liposomes encapsulating DTPA in rats injected with a soluble form of Pu which mimics the partitioning of dissolved Pu following its absorption into the blood from the primary site of contamination (lung or wound site depending on the route contamination).

2 Materials and Methods

Unilamellar conventional liposomes (DSPC/Cholesterol/DSPE 69:30:1) were used to encapsulate a solution of DTPA as Na_3Ca-DTPA (25mM).

Rats were contaminated by intravenous administration of the soluble citrate form of ^{238}Pu (7 to 10.3kBq) and received or not an injection of DTPA either as marketed free form or entrapped within liposomes (5 µmol/kg). Treatment schedules started at one hour (prompt treatment) or at seven days (delayed treatment) after contamination and were given as a single injection or repeated injections. A prophylactic single treatment was also tested given at three days before Pu exposure.

* Corresponding author : olivier.gremy@cea.fr

Excreta were collected and organs of interest (liver, bone and spleen mainly) were excised at euthanasia for measurement of residual Pu alpha activity. The comparative effects of liposomal and free DTPA at similar doses were examined in terms of fecal and urinary Pu overexcretion and limitation of Pu burden in tissues from rats receiving the various treatment regimens.

3 Results

The prompt treatment (1 hour) with liposomal DTPA at 5μmol.kg^{-1} prevented significantly by about 75%, 52%, and 77% Pu deposition in liver, bone and spleen as compared to controls, respectively. Thus, this galenic form of DTPA reduced Pu retention by 2.5-, 1.6- and 3-fold in liver, bone and spleen, respectively, compared with free DTPA at similar dose. Similar to data for tissues, liposomal DTPA had a significantly greater effect on urinary elimination of Pu compared with free DTPA during at least one week after prompt treatment. Whereas Pu level excretion in feces increased in the order no treatment < free DTPA < liposomal DTPA at early times, this order was reversed at later times. The total fecal excretion does not seem to be enhanced by DTPA treatments but only accelerated compared to no treatment, and in a faster way with liposomal DTPA than with free DTPA.

For delayed administration (7 days), liposome-entrapped DTPA decreased hepatic, skeletal and splenic Pu levels by, respectively 68, 39 and 74% of the levels in untreated control rats (only 19, 12 and 24% with a similar dose of free DTPA). Concerning urinary and fecal Pu excretions after delayed DTPA treatment, similar observations as those referred to the early treatment have been made.

In addition, four injections of liposomal DTPA begun at 7 days and spaced at 3-day intervals further improved the removal of Pu especially from bone (-55% as compared to untreated rats) compared to a single injection (-39% as compared to untreated rats).

Lastly, a treatment by liposomal DTPA given three days before Pu exposure prevented Pu accumulation at seven days by about 47%, 29% and 63% in the liver, bone and the spleen, respectively, compared to untreated control rats, which shows an efficacy of a prophylactic treatment with liposomal-encapsulated DTPA.

4 Discussion/Conclusion

DTPA formulated in our liposomes can significantly improve Pu decorporation from systemic tissues regardless of the treatment regimen applied.

As the expected fate of liposomes is their capture by phagocytic cells in organs of the mononuclear phagocytic system, the higher effectiveness of liposomal DTPA in removing Pu from the liver and the spleen, as compared to free DTPA, was undoubtedly related in part to a far greater accumulation of DTPA molecules in phagocytic cells of these tissues. However, the early increased Pu excretion in feces following liposomal DTPA treatment suggests that intracellular chelation of Pu can also take place in other cells than phagocytic cells, such as hepatocytes. The success of a prophylactic treatment with liposomal DTPA given three days prior Pu exposure shows a persistence of DTPA molecules in tissues over several days, thereby providing another advantage to liposome-encapsulated DTPA over free DTPA in chelating intracellular Pu.

Thanks to this high and persistent accumulation of DTPA in soft tissues when delivered as liposomal form, these tissues certainly act as reservoirs for DTPA. DTPA molecules that are slowly released from them into the blood stream may largely contribute to a sustained extracellular action of DTPA on Pu still deposited on bone surfaces, contrary to free DTPA which has a very short blood half-life. A contribution of DTPA molecules slowly lost from liposomes during their blood circulation to the bone Pu decorporation is not excluded but it is assumed to be marginal because of the high stability of our liposomes.

As a consequence, the advantage of liposomal DTPA over marketed free DTPA would be not only directly but also indirectly due to the better cell penetration of DTPA when loaded within liposomes.

To conclude, encapsulation of DTPA in suitable liposomes is therefore a good therapeutic strategy for improving delivery, and thereby enhancing decorporation efficacy for transuranic actinides such as plutonium. As compared to free DTPA, this galenic formulation could be advised for both early and delayed chelation therapy. Besides, the sustained action of DTPA when encapsulated within liposomes suggests a great efficacy of a treatment regimen consisting of spaced repeated administrations.

Results presented here were obtained in the frame of the Orano-Inserm-CEA collaboration.

Modelling DTPA decorporation of Am in rats

Manuel Kastl[1,2*], *Olivier* Grémy[3], *Laurent* Miccoli[3], *Stéphanie* Lamart[3], *Weibo* Li[1], and *Augusto* Giussani[4]

[1]Helmholtz Zentrum München, German Research Center for Environmental Health, 85764 Neuherberg, Germany
[2]Technische Universität München, Fakultät für Physik, 85748 Garching, Germany
[3]Laboratoire de Radiotoxicologie CEA/LRT, Université de Paris-Saclay, Bruyères-le-Châtel, France
[4]BfS – Federal Office for Radiation Protection – Department of Radiation and Health, 85764 Oberschleißheim, Germany

Decorporation therapy with DTPA (diethylene triamine pentaacetic acid) is recommended as a standard procedure after accidental intake of transuranics such as plutonium/americium (Pu/Am). DTPA is able to chelate the incorporated Pu/Am and eliminate them rapidly from the body, mainly by enhancing their urinary excretion rate. In general Ca/Zn-DTPA is administered following empirical protocols. Since the chelation process perturbs the regular biokinetics of Pu/Am, standard biokinetic models are inadequate for interpreting monitoring data and estimating dose to the patient. Moreover, optimization of the treatment strategy and prospective assessments of the therapy success in terms of averted dose cannot be evaluated by using the current models.

In the recent years a European collaboration within the frame of EURADOS (European Radiation Dosimetry Network) has been initiated to develop a compartmental model approach to this issue. The basic idea of the adopted approach is to consider the biokinetics of Pu/Am and of the injected DTPA separately and to couple them by a suitable mathematical description of the chelation mechanism as a second-order process [1, 2]. Except the proposal presented by Konzen and Brey [3], which is based on the EURADOS approach, all other attempts existing in the literature [4-9] are empirical and developed mainly for the interpretation of one or a limited number of specific incorporation cases.

In this work the available experimental data from controlled animal studies with Am recently published by Grémy et al. [10] was used to define the biological sites of chelation and the chelation rate constants. Rats were given DTPA either prior to or after Am intravenous injection. In these studies all boundary conditions are precisely known, differently from the case of accidental intakes in humans. The chelation rate constants and the identification of the chelation sites are thus the only unknown variables of the system.

In a first step, chelation was assumed to occur only in extracellular fluids, with two different rate constants for chelation in blood and chelation in other fluids (lymph, interstitial fluids). Model predictions reproduce the accumulation of activity in urine until time of sample collection, corresponding to a 24h period in the studies of Grémy et al. Predictions were compared with the experimental 24h urinary excretions (Fig. 1). While there is a good agreement between prediction and data when DTPA has been administered shortly after contamination (top left panel), the agreement for the other experiments is less satisfactory. Moreover, increased faecal excretion and decorporation from liver and

* Corresponding author: manuel.kastl@helmholtz-muenchen.de

skeleton cannot be reproduced under these simple assumptions. For this reason, further simulations are planned, considering in addition intracellular chelation.

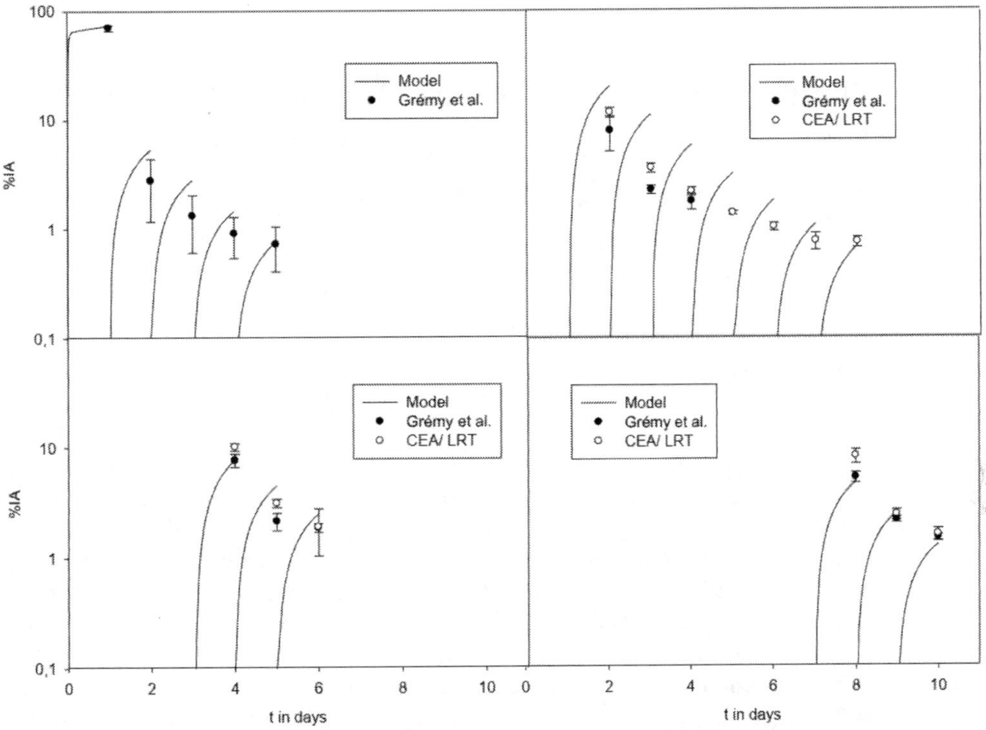

Figure 1. Comparison of measured 24h urine excretion with the corresponding model predictions for single DTPA injection at different times after contamination: 3 minutes (top left), one day (top right), three days (bottom left) and seven days (bottom right). Values are expressed as the percentage of the initial Injected Activity (mean ± sd (% IA)). "Grémy et al." are data presented in [10]; "CEA/LRT" are unpublished data from the same laboratory.

References

1. B. Breustedt, E. Blanchardon, P. Berard, et al., Radiat. Prot. Dosim. **134** 1 (2009)
2. M. Kastl, A. Giussani, E. Blanchardon, B. Breustedt, P. Fritsch, C. Hoeschen, M.A. Lopez, Int. J. Radiat. Biol. **90** (2014)
3. K. Konzen and R. Brey, Health Phys. **108** (2015)
4. R.R. Bailey, K.F. Eckerman, L.W. Townsend, Radiat. Prot. Dosim. **105** (2003)
5. P. Fritsch, L. Grappin, A.M. Guillermin, R. Fottorino, M. Ruffin, A. Mièle, Radiat. Prot. Dosim. **127** (2007)
6. R.M. Hall, G.A. Poda, R.R. Fleming, J.A. Smith, Health Phys. **34** (1978)
7. A.C. James, L.B. Sasser, D.B. Stuit, S.E. Glover, E.H. Carbaugh, Radiat. Prot. Dosim. **127** (2007)
8. J.J. Jech, B.V. Andersen, K.R. Heid, Health Phys. **22** (1972)
9. T.R. LaBone, Internal Radiation Dosimetry (Medical Physics Publishing, 1994)
10. O. Grémy, D. Laurent, S. Coudert, N.M. Griffiths, L. Miccoli, Rad. Res. **185** (2016)

Solid-liquid exchange between uranium and a synthetic apatite: towards uranium decorporation from bone matrix

Jingxian Wang[1,*], *Damien* Bourgeois[1] , and *Daniel* Meyer[1]

[1]ICSM, LHYS, BP 17171, 30207 Bagnols-sur-Cèze, France

Abstract

Natural uranium exhibits chemical toxicity, especially known with its acute effects on kidney. Simultaneously, it has been proved that uranium accumulates in bones during long-term exposure[1] but its chronical effects on bones are not clear. Particularly the mechanisms associated to accumulation into and release from bones are unknown, which is key to design and test decorporation reagents in future. Bone is a complicated organ, composed of mineralized apatite and organic compounds (mostly type I collagen). Our work is dedicated to the understanding of how uranium is accumulated in the inorganic bone matrix through chemical pathways, and what factors influence the solid/liquid equilibrium between uranium and the bone. To fulfil this goal, apatite materials which mimic bone apatite have been synthesized, with and without uranium. Such apatite materials have been pre-equilibrated with a fluid mimicking blood plasma at physiological pH=7.4.

The proposed exchange mechanism of +2 cations (Ca^{2+}, Zn^{2+}, etc.) with bone relies on a two compartment model[2], including ion exchange in the amorphous hydration shell and further ion diffusion into the mineral matrix. From our previous work and other researchers[3-6], bone apatite has high affinity to uranium. However, the two compartment model has not been applied to U exchange, and studies are complicated by the complex speciation of uranium in vivo. U(VI) in physiological fluids forms many inorganic complexes, such as with carbonate and calcium, and organic complexes, instead of "free" $\{UO_2^{2+}\}$[7]. In our study, U(VI) is introduced in liquid containing carbonate ions in order to investigate the exchanges of U(VI)-carbonato complexes with the synthetic apatite stated above. Both the thermodynamics and the kinetics behaviors have been analyzed, and the exchange reversibility is further investigated. The apatite matrix after interaction with uranium is also characterized with various techniques, including electron microscopy (TEM and SEM) with energy dispersive X-ray spectroscopy, X-ray diffraction and Raman spectroscopy, to locate uranium and specify possible uranium incorporated phases. Furthermore, X-ray adsorption technique has been applied to study the uranium coordination mode on surface and in mineral, especially the uranium-containing complex specification. Finally, a solid/liquid exchange model is set and it will be employed to evaluate the influence of U(VI) chelators in liquid phase on solid/liquid exchange process, towards to further uranium detoxification studies.

* Corresponding author: jingxian.wang@cea.fr

References

1. C. Vidaud, D. Bourgeois, D. Meyer, Chem. Res. Toxicol. **25**, 1161 (2012)
2. V.E. Badillo-Almaraz, J. Ly, J. Colloid Interface Sci. **258**, 27 (2003)
3. D. Bourgeois, B. Burt-Pichat, X. Le Goff, J. Garrevoet, P. Tack, G. Falkenberg, L. Van Hoorebeke, L. Vincze, M.A. Denecke, D. Meyer, C. Vidaud, G. Boivin, Anal. Bioanal. Chem **407**, 6619 (2015)
4. G. Chatelain, D. Bourgeois, J. Ravaux, O. Averseng, C. Vidaud, D. Meyer, J. Biol. Inorg. Chem. **20**, 497 (2015)
5. P. Thakur, R.C. Moore, G.R. Choppin, Radiochim. Acta **93**, 385 (2005)
6. C.C. Fuller, J.R. Bargar, J.A. Davis, Environ. Sci. Technol. **37**, 4642 (2003)
7. G. Bernhard, G. Geipel, T. Reich, V. Brendler, S. Amayri, H. Nitsche, Radiochim. Acta **89**, 511 (2001)

BIO Web of Conferences **14**, 06005 (2019) https://doi.org/10.1051/bioconf/20191406005
HEIR 2018

Uranyl-chelating peptides to help understanding uranium toxicity at a molecular level

Fanny Laporte,[1] *Matthieu* Stark,[1] *Colette* Lebrun,[1] *Stéphane* Oros,[1] *Nathalie* Sisommay,[1] *Vicky* Gathu,[1] *Pier Lorenzo* Solari,[5] *Gaëlle* Creff,[3] *Jérôme* Roques,[4] *Christophe* Den Auwer,[3] *Claude* Vidaud[2] and *Pascale* Delangle[1,*]

[1]INAC-SyMMES, Univ. Grenoble Alpes, CEA, CNRS, 38000 Grenoble, France
[2]BIAM, CEA, CNRS, Aix-Marseille Université, 13108 Saint-Paul-lez-Durance, France
[3]Institut de Chimie de Nice, Université Côte d'Azur, 06108 Nice, France
[4]Institut de Physique Nucléaire d'Orsay, CNRS-IN2P3, Univ. Paris-Sud, Univ. Paris-Saclay, France
[5]Synchrotron SOLEIL, L'orme des Merisiers, Saint-Aubin, 91192 Gif-sur-Yvette, France

Context

Uranium is a natural element widely found in the environment, due to both natural occurrence in mineral ores or in sea water and industrial applications. The production of nuclear energy uses enriched uranium in ^{235}U for nuclear fission. Despite its ubiquitous distribution, uranium has no essential role in living organisms and presents radiological and chemical toxicities. Despite significant recent advances in the field, there is still a serious lack of knowledge about the molecular interactions responsible for uranium toxicity. The underlying mechanisms need to be unravelled to predict the effect of uranium on living organisms and also to help in designing efficient detoxification agents, to be used in case of dirty bombs or accidental release of uranium in the environment.

Design of biomimetic peptides

At the CIBEST lab, we propose a biomimetic approach to help understanding the interactions of metals with biological molecules, relevant to their regulation and toxicity.[1, 2] This strategy was applied for uranium, which most stable form *in vivo* is the uranyl cation, UO_2^{2+}, which prefers four to six oxygen donors in its equatorial plane, perpendicular to O-U-O bonds. Short peptide sequences are used as simple models of metal-binding sites in proteins to get structural and thermodynamic data, which are not accessible directly with large biological molecules. Cyclic constrained peptide scaffolds that orient metal-coordinating groups found in proteins toward the equatorial plane of the uranyl cation revealed to be highly effective uranyl-binding peptides.[3]

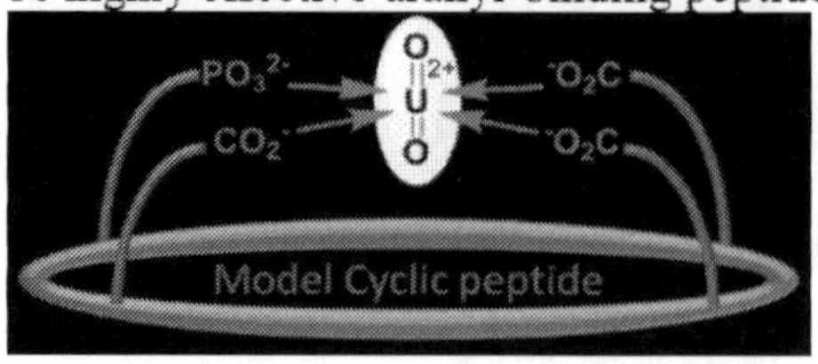

* Corresponding author: pascale.delangle@cea.fr

Results

A combination of synthesis, analytical chemistry, X-Absorption Spectroscopy and theoretical calculations allowed us to correlate the amino acid sequence to the stability of the uranyl complexes. Uranium-coordinating amino acids with side-chains containing oxygen donors, such as aspartic acids, glutamic acids and phosphoserines were inserted in the cyclic peptide sequences. The largest uranyl complexes stabilities were obtained when four coordinating amino acids were converging to the uranyl equatorial plane.[3-5]
Importantly, phosphate groups significantly enhance the affinity of the biomimetic peptides for uranyl, which strongly supports the importance of phosphoamino acids in uranyl binding in proteins. Therefore, our biomimetic approach reinforces the relevance of considering phosphoproteins such as osteopontin as potential uranyl targets in vivo.[6-8]

Acknowledgements

This work was supported by the toxicology and NRBC programs of CEA and the LabEx ARCANE (Grant ANR-11-LABX-0003-01).

References

1. O. Sénèque, P. Rousselot-Pailley, A.M. Pujol, D. Boturyn, S. Crouzy, O. Proux, A. Manceau, C. Lebrun, P. Delangle, Inorg. Chem., **57**, 2705 (2018)
2. A.S. Jullien, C. Gateau, C. Lebrun, P. Delangle, Inorg. Chem., **54**, 2339 (2015)
3. C. Lebrun, M. Starck, V. Gathu, Y. Chenavier, P. Delangle, Chem. Eur. J., **20**, 16566 (2014)
4. M. Starck, F. A. Laporte, S. Oros, N. Sisommay, V. Gathu, P. L. Solari, G. Creff, J. Roques, C. Den Auwer, C. Lebrun, P. Delangle, Chem. Eur. J., **23**, 5288 (2017)
5. M. Starck, N. Sisommay, F. A. Laporte, S. Oros, C. Lebrun, P. Delangle, Inorg. Chem., **24**, 11557 (2015)
6. L. Qi, C. Basset, O. Averseng, E. Quemeneur, A. Hagege, C. Vidaud, Metallomics, **6**, 166 (2014)
7. T. N. S. Huynh, C. Vidaud, A. Hagege, Metallomics, **8**, 1185 (2016).
8. F. Laporte, PhD thesis, Univ. Grenoble Alpes (2017).

Pharmacological study of stable potassium iodide (KI) repeated prophylaxis in adult rats

Guillaume Phan[1,*], *François* Rebière[1], *Rym* Chioukh[1], *David* Suhard[1], *Alexandre* Legrand[1], *Floriane* Carpentier[1], *Charlotte* Moulin[1], *Thibaud* Sontag[1], *Maâmar* Souidi[1], *Céline* Bouvier-Capely[1], *Jean-René* Jourdain[2], *Michelle* Agarande[3], and *Valérie* Renaud-Salis[4]

[1]IRSN, PSE-SANTE/SESANE, 92260 Fontenay-aux-Roses, France

[2]IRSN, DAI, 92260 Fontenay-aux-Roses, France

[3]IRSN, PSE-ENV/SAME, 78110 Le Vésinet, France

[4]IRSN, PSE-SANTE/SESUC, 92260 Fontenay-aux-Roses, France

1 Introduction

Ingestion of stable iodine, mainly as potassium iodide (KI) tablets, is the only medical countermeasure available to prevent the binding of radioactive isotopes of iodine to the thyroid and to protect the gland from the occurrence of cancer. The international guidelines recommend the administration of a single dose of 130 mg KI in adults, corresponding to approximately 1.8 mg/kg, and envisage the possibility of a second dose in case of the impossibility to evacuate the populations. However, the nuclear reactors accidents at Fukushima in 2011 showed that the populations could be exposed to prolonged or repeated releases of radioisotopes of iodine for more than one week. In such situations, prolonged KI prophylaxis could be considered but no indication is currently available regarding the conditions for implementing repeated doses of stable iodine. The aim of this work was to study the protective effect of different KI doses in the rats in order to select an optimal KI dose which could be implemented in repeated prophylaxis. The pharmacokinetic parameters of stable iodine were then evaluated in the rats after oral ingestion of KI in order to model and assess a dose regimen for a repeated prophylaxis over 8 days.

2 Materials and methods

In a dose-response study, the effect of different oral single doses of KI (0.001; 0.01; 0.1; 1; 1.8; 5 mg/kg) was evaluated in the male Wistar rats (n=6 per dose) in terms of protection of the thyroid against the incorporation of radioactive iodine I-125 (1.11 MBq/kg) injected intravenously 1 hour after KI dosing. The thyroids were dissected 1 hour after exposure to I-125 and the total radioactivity of I-125 was determined by gamma counting (multidetector Packard).The plasma stable iodine concentration as a function of the time after oral single dose of 1 mg/kg KI was measured by inductively coupled plasma mass spectrometry (ICP-MS X Series II and iCAP Q, Thermo Electron) (n=6 per time point). The pharmacokinetic parameters of stable iodine were determined and the dose regimens were simulated using

* Corresponding author: guillaume.phan@irsn.fr

Phoenix WinNonlin v.6.3 software (Certara). The pharmacodynamic and pharmacokinetic modelling were performed using Phoenix WinNonlin v.6.3 and Prism v.5.0.1 (GraphPad) softwares. A dose regimen consisting of 1 mg/kg daily during 8 days was tested in rats (n=4 per time point) and verified through iodine plasma measurements at different times during the treatment (at 3; 5; 9 and 10 days).

3 Results and discussion

The protection of the thyroid in terms of avoided iodine I-125 activity compared with control rats as a function of KI dose is described by a saturation-type function (Figure 1). The results showed that avoided activities above 85% could be obtained for KI doses higher than 1 mg/kg. This dose could be hence selected as a minimum effective KI dose for repeated prophylaxis [1, 2].

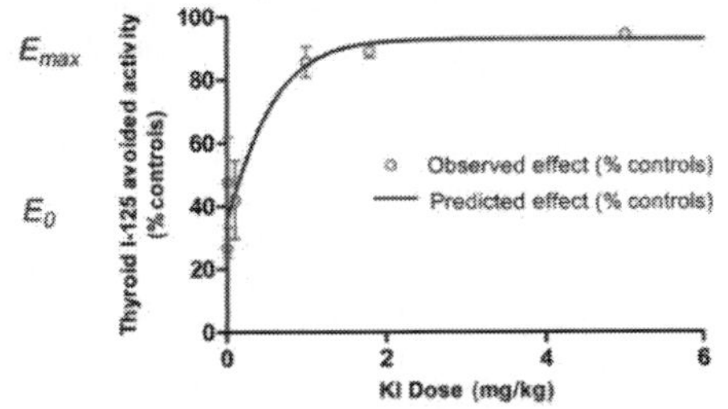

Fig.1. Stimulation model for iodine I-125 avoided activity in the thyroid as a function of stable KI dose (E_0: initial effect; E_{max}: maximum effect)

A correlation between plasma stable iodine I-127 and avoided I-125 activity in the thyroid was observed and showed that protection efficiencies from 63% to 88% could be achieved during 24h after oral administration of 1 mg/kg KI (data not shown). A dose regimen consisting of 1 mg/kg daily was modelled and tested in rats for 8 days. The plasma measurements at different times showed that stable iodine I-127 concentrations were consistent with the theoretical profile (Figure 2). This dose regimen should ensure high thyroid protection efficiencies throughout the treatment and could be proposed as an effective dose regimen for repeated KI prophylaxis [3].

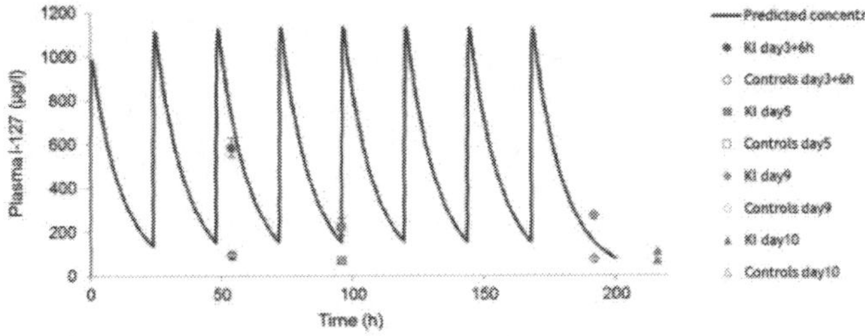

Fig.2. Plasma iodine concentrations measured at different times in rats treated by 1 mg/kg KI daily for eight days.

Acknowledgements
These studies were supported by the research program in Nuclear Safety and Radiation Protection of the French National Research Agency (ANR) (contract number ANR-11-RSNR-0019)

References

[1] G. Phan, F. Rebiere, D. Suhard, A. Legrand, F. Carpentier, T. Sontag, M. Souidi, J.R. Jourdain, M. Agarande, V. Renaud-Salis, Optimal KI Prophylactic Dose Determination for Thyroid Radiation Protection After a Single Administration in Adult Rats, Dose-response : a publication of International Hormesis Society 15(4) (2017) 1-8.

[2] M. Benderitter, T. Pourcher, J.-C. Martin, J. Darcourt, P. Guigon, F. Caire-Maurisier, A. Pech, D. Lebsir, C. Rosique, J. Guglielmi, F. Rebière, K. Tack, G. Phan, P. Lestaevel, M. Souidi, J.-R. Jourdain, Do multiple administrations of stable iodine protect population chronically exposed to radioactive iodine: what is PRIODAC research program (2014–22) teaching us?, Radiation Protection Dosimetry (2018) ncy129.

[3] G. Phan, R. Chioukh, D. Suhard, A. Legrand, C. Moulin, T. Sontag, F. Rebière, C. Bouvier-Capely, M. Agarande, V. Renaud-Salis, J.-R. Jourdain, Repeated KI Prophylaxis in Case of Prolonged Exposure to Iodine Radioisotopes: Pharmacokinetic Studies in Adult Rats, Pharmaceutical Research 35(12) (2018) 227.

Impact of repeated dose of stable iodine in an *in utero* rat model using a metabolomic approach

Clément Rosique[1] , *Dalila* Lebsir[2] , *Maâmar* Souidi[2] , and *Jean-Charles* Martin[1]

[1]Aix Marseille Univ, INSERM, INRA, C2VN, Marseille, France
[2]Institut de Radioprotection et de Sûreté Nucléaire (IRSN), PRP-HOM, SRBE, LRTOX, 92260 Fontenay-aux-Roses, France

Abstract

The Fukushima nuclear power plant blast resulted in the release of ^{131}Iodine for several weeks. This unexpected issue challenged the iodin doctrine [1], in which the counter-measure is to provide a unique iodine tablet to saturate thyroid during the radioactive contamination not expected to last more than several hours. A new doctrine must be implemented to take into account such case of extended exposure based on repeated iodine administration with adapted dosage. But repeated administration of iodine can block the thyroid [2] and few scientific evidences regarding repeated iodine administration (and its potential undesirable effect) are at our disposal [3]. Moreover, unborn and young children are at high risk during a nuclear incident: it is currently recognized that one of the risks of exposure to radioactive iodine is the development of thyroid cancer, especially when exposure occurred during childhood [4]. Their protection is a main priority. Our goal was to evaluate the potential undesirable effects of such repeated iodine administration in the offspring using an untargeted metabolomic approach on a rat reproductive model.

Pregnant rats received repeated doses of potassium iodine (KI group: 1mg/kg/24h) or water for injection (control group) for 8 days. The potential metabolic disruption was investigated in the offspring 30 days after weaning.

Using LC-MS, we compared the blood's metabolite composition between KI and control male rats; using a high throughput annotation procedure with an in-house databank, 264 metabolites were annotated (based on retention time and exact mass), combined in 52 functional biological modules/pathways, and converted into corresponding scores using a PLS multiblock algorithm (see Fig. 1).

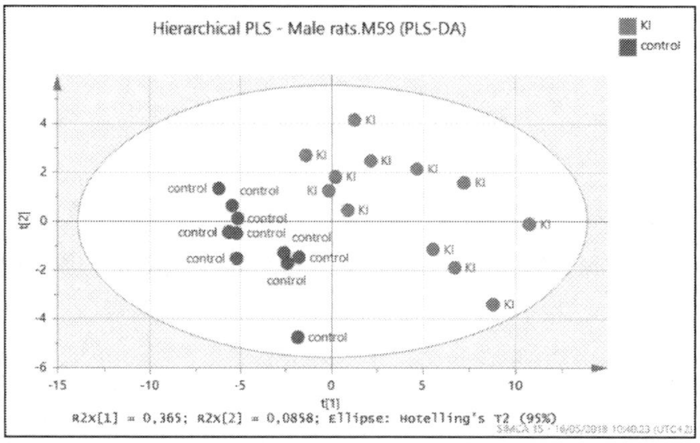

Fig. 1. Hierarchical PLS-DA based on 51 functional biological modules, CV-ANOVA p-value = 0.0018

Running a random forests test, we found 19 modules significantly impacted by the KI treatment (VIP values > 0.05) including pathways of redox status, aminoacids, TCA cycle or oxidative stress. These findings indicated a prenatal effect of KI administration that lasted over the long term (adolescence). Whether or not these outcomes are pathological is unknown.

References

[1] World Health Organization, « The World Health Report 1999 » (1999).

[2] S. Dreger, M. Pfinder, L. Christianson, S. K. Lhachimi, et H. Zeeb, « The effects of iodine blocking following nuclear accidents on thyroid cancer, hypothyroidism, and benign thyroid nodules: design of a systematic review », *Systematic Reviews*, vol. **4**, n° 1 (2015).

[3] P. Mitrou, S. A. Raptis, et G. Dimitriadis, « Thyroid disease in older people », *Maturitas*, vol. 70, n° **1**, p. 5‑ 9 (2011).

[4] E. K. Cahoon *et al.*, « Risk of Thyroid Nodules in Residents of Belarus Exposed to Chernobyl Fallout as Children and Adolescents », *The Journal of Clinical Endocrinology & Metabolism*, vol. 102, n° **7**, p. 2207‑ 2217 (2017).

Chelating properties of the calixarene carboxylic

Pascal Chanton[1]

Laboratoire CEVIDRA, France

IRSN demonstrated that a nanoemulsion loaded with calixarene carboxylic is much more efficient for the decontamination of skin contaminated by actinides than the existing products [1, 2, 3, 4]. This innovation has been patented by IRSN and an exploitation license has been granted to a Pharmaceutical Laboratory

The efficacy of the IRSN product is based on the known complexation activity of the calixarene carboxylic on actinides, and on integrating this chelating agent in a nanoemulsion.

A cleansing nanoemulsion formulation has been developed by IRSN on the advice of the physicians of the nuclear industry. This cleansing formulation of calixarene carboxylic demonstrated a similar efficacy to the patented nanoemulsion [5].

This cleansing nanoemulsion has been developed to be used as an emergency treatment at the time of actinides contamination or potential contamination, the product acting by chelating the soluble elements of actinides and by cleansing insoluble elements present on the skin/hairs.

An industrial cleansing nanoemulsion has been made and marketed in compliance with the IRSN one. It was therefore important to better determine the chelating properties of the calixarene carboxylic when loaded in the industrial cleansing nanoemulsion to better define the scope of use of this marketed product.

Summary

The Calixarene carboxylic is a the p-tert-butylcalix[6]arene bearing 3 carboxylic groups arranged in C3 symmetry with a cone conformation allowing the formation of a pseudo planar hexacoordinate complex with Uranium, displaying a high affinity (6).

This property was initially used by IRSN for the analysis of contaminated urine / water. The use of calixarene columns (LCarbH3 and LHydH3) allowed a quantitative and reproducible separation of Uranium, Plutonium and Americium from mineralized urine samples (7). Thorium extraction from mineralised urine samples on calixarene columns was feasible and demonstrated with quantitative yields (8).

The chelating properties of the calixarene carboxylic in a nano emulsion was assessed first by performing the "In vitro" study developed by IRSN. In a second step, IRSN demonstrated that specific chelating capacities of the nano emulsion allowed an efficient decontamination of skin using an "Ex-vivo" study model.

The "In vitro" study performed by IRSN against Uranium, Cobalt, Caesium and Strontium shows that the Calixarene CEVIDRA's cleansing nano emulsion exhibits good

[1] Corresponding author: p.chanton@cevidra.com

affinity towards actinides such as Uranium, but also towards other elements such as Cobalt, Strontium or Caesium.

Compared to previous studies, a relatively lower extraction efficacy regarding Uranium can be observed. This would be potentially due to a one step extraction of 4 mixed elements instead of one element extraction in previous studies. This one step extraction of 4 mixed elements could also induce competition between Uranium and the other "lighter" and more diffusible elements

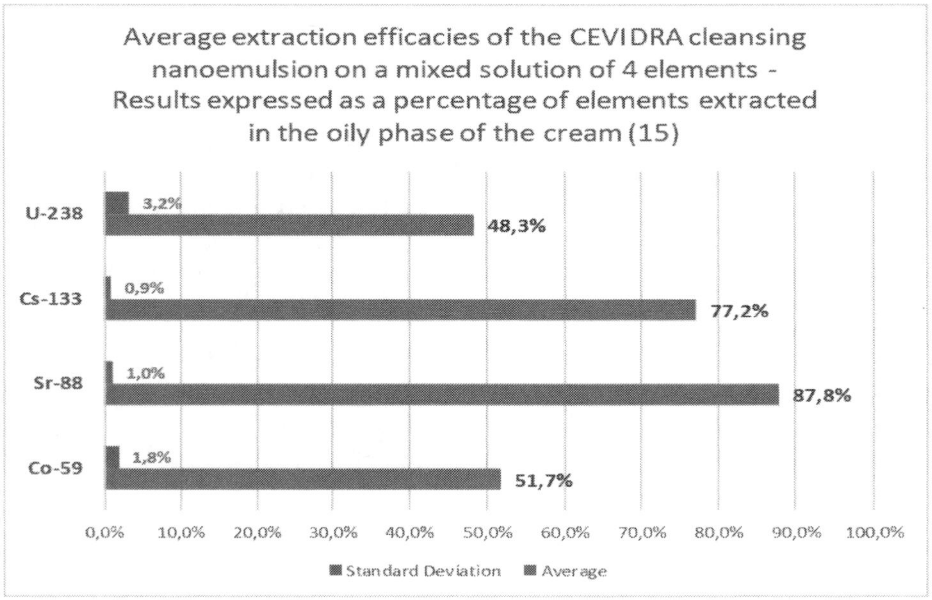

These results allow CEVIDRA to extend the use of the product to the treatment of Cobalt, Caesium and/or Strontium skin and hair contaminations.

The safety studies performed on the active ingredient and performed on the CEVIDRA's cleansing nano emulsion, demonstrated the good tolerance and the very low toxicity of this product.

Recently, the use of the CEVIDRA nano emulsion on real cases of skin and hair contaminations confirmed the high decontamination efficacy of the product.

These elements reinforce the importance of using the CEVIDRA nano emulsion for the emergency treatment of contamination or suspicion of contamination to Actinides (Uranium, Plutonium, Americium and Thorium), Cobalt, Caesium and Strontium.

References

1. Spagnul, A., Bouvier-Capely, C., Phan, G., Rebière, F. et Fattal, E. (2010). "Calixarene-entrapped nanoemulsion for uranium extraction from contaminated solutions". Journal of Pharmaceutical Sciences, 99(3), 1375-1383
2. Spagnul, A., Bouvier-Capely, C., Adam, M., Phan, G., Rebière, F. et Fattal, E. (2010). "Quick and efficient extraction of uranium from a contaminated solution by a calixarene nanoemulsion". International Journal of Pharmaceutics, 398(1-2), 179-184
3. Spagnul, A., Bouvier-Capely, C., Phan, G., Landon, G., Tessier, C., Suhard, D., Rebière, F., Agarande, M. et Fattal, E. (2011). "Ex vivo decrease in uranium diffusion

through intact and excoriated pig ear skin by a calixarene nanoemulsion". European Journal of Pharmaceutics and Biopharmaceutics, 79(2), 258-267

4. Grives, S., Phan, G., Morat, G., Suhard, D., Rebiere, F. et Fattal, E. (2015). "Ex vivo uranium decontamination efficiency on wounded skin and in vitro skin toxicity of a calixarene-loaded nanoemulsion". Journal of Pharmaceutical Sciences, 104(6), 2008-2017
5. Phan, G., Semili, N., Bouvier-Capely, C., Landon, G., Mekhloufi, G., Huang, N., Rebière, F., Agarande, M. et Fattal, E. (2013). "Calixarene cleansing formulation for uranium skin contamination". Health Physics, 105(4), 382-389
6. Fattal E, Tsapis N, Phan G. Novel drug delivery systems for actinides (uranium and plutonium) decontamination agents. Adv Drug Deliv Rev. 2015 Aug 1;90:40-54.
7. Bouvier-Capely, C., Manoury, A., Legrand, A., Bonthonneau, J. P., Cuenot, F. et Rebière, F. (2009). "The use of calix[6]arene molecules for actinides analysis in urine and drinking water: An alternative to current procedures". Journal of Radioanalytical and Nuclear Chemistry, 282(2), 611-615
8. Mekki, S., Bouvier-Capely, C., Jalouali, R. et Rebière, F. (2010). " The extraction of thorium by calix[6]arene columns for urine analysis ". Radiation Protection Dosimetry, 144(1-4), 330-334

BIO Web of Conferences **14**, 06009 (2019) https://doi.org/10.1051/bioconf/20191406009
HEIR 2018

Decorporation therapy. The "Precautionary approach" versus the "Urgent approach"

Alexis Rump[1], *Patrick* Ostheim[1,*], *Andreas* Lamkowski[1], *Stefan* Eder[1], *Michael* Abend[1], and *Matthias* Port[1]

[1] Bundeswehr Institute of Radiobiology affiliated to the University of Ulm, Munich, Germany

1 Background

In case of a nuclear accident or a terrorist attack by a „dirty bomb“, there is a risk of contamination and incorporation of radionuclides besides external irradiation. A treatment by chelators and adsorptive agents is particularly effective when started early for pharmacokinetic reasons. Starting decorporation therapy without knowing the committed effective dose may however unnecessarily expose the patient to the side-effects of the medication. The question at issue is:

1. Delay the therapy to wait for the results of internal dosimetry?
2. Start the therapy promptly on spec?

2 Method

Selective review of the literature.

3 Results

Many authors stress the importance of the committed effective dose to decide on the necessity of decorporation therapy ("precautionary approach"). It is acknowledged that therapy is more effective when started early and that internal dosimetry takes time. But excepted for iodide, clear cut rules on when to start the treatment are not formulated. A Canadian defense source states that starting decorporation without a previous dose assessment requires a case by case decision. On the other side, official French guidelines recommend to start treatment "à priori" within 2 hours after a suspected incorporation.

*corresponding author: patrickostheim@bundeswehr.org

4 Conclusion/Prospect

Excepted for iodide there is no internationally accepted standard on when to start a decorporation therapy. In consideration of the higher effectiveness and the low side-effects of a short-term treatment, it seems reasonable to start decorporation therapy as soon as possible, even before the committed effective dose has been assessed. The decision of continuation or discontinuation of the therapy should be taken after internal dosimetry is completed on the basis of the committed effective dose.

Evaluation Of Protracted Chelation Treatments To Decorporate Plutonium After Rat Lung Contamination

Laurent Miccoli[1], *Sylvie* Coudert[1], *Pierre* Laroche[2], and *Olivier* Grémy[1]

[1]Laboratoire de RadioToxicologie, CEA, Université de Paris-Saclay, Bruyères-le-Châtel, France
[2]Orano, Courbevoie, France

In the electronuclear industry, accidental release of alpha-emitting transuranic actinides such as plutonium (Pu) may occur at the work place. As a consequence, internal contamination *via* inhalation is a potential hazard for nuclear workers handling Pu. After deposition in the lungs, the transportable fraction of Pu translocates from the lungs to the bloodstream. The importance of this fraction and its translocation kinetic depend on numerous variables, including the physicochemical form of the inhaled Pu compound. Although a small fraction of the systemic burden is cleared in urine, most of the transferred activity is retained in extrapulmonary tissues, mainly in liver and bones [1].
The only medical approach available nowadays for reducing the committed radiological dose in target tissues, and consequently the risk of radiation-induced pathologies, is the removal of Pu from the body by chelation therapy with diethylenetriaminepentaacetic acid (DTPA used as Ca or Zn trisodium salt) [2]. Unfortunately, DTPA has serious drawbacks because of a very rapid elimination from the body and a poor distribution in Pu retention sites [3]. The success of a DTPA treatment in removing internalized Pu varies greatly, depending not only on its physicochemical form [4], but also on the treatment methods and regimens used (i.e., dose level, single or repeated administrations, frequency, administration route, and delay post-exposure). Once deposited in target tissues Pu becomes very difficult to remove. Thus, repeated administrations of DTPA which are more effective than a single one are needed [5].
For a decorporating treatment based on multiple DTPA injections, several therapeutic variables should be considered such as the number and the frequency of injections, or the dose level per injection. However, even though various data are available in the literature, they remain fragmentary and not sufficient, in our opinion, to draw definitive general conclusions for an optimal protracted chelation therapy.
In this context, the aim of the present work was to better understand the influence of important therapeutic variables on the overall efficacy of a repeated DTPA chelation treatment. This new useful information may help to design better treatment schedules that can be both efficient and the least burdensome as possible for the patient. Thus, our experimental work design featured: (1) all DTPA treatment schedules commenced several days or weeks following contamination in order to consider only Pu firmly-fixed in tissues and not the available pool of circulating Pu just after contamination; (2) the various DTPA treatment regimens tested used comparable cumulated total dose to assess only the effects of the distribution and/or the frequency of multiple injections as therapeutic variables; (3)

the effects of the variables on the efficacy were evaluated in both systemic tissues and lungs.
We here designed animal experiments to assess mainly the impact of the fractionation of a given dose and that their frequency totaling the same dose. Firstly, rats were contaminated intravenously with Pu then treated several days later with DTPA given at once or as various split-dose regimens cumulating to the same total dose and spread over several days. Similar efficacies were induced by the injection of the total dose or by splitting the dose in several smaller doses independently of the number of doses and the dose level per injection. Secondly, rats were pulmonary contaminated with a moderately form of Pu nitrate then received three weeks later a DTPA dose eleven-fold higher than the maximal daily recommended dose, administered either as a single bolus or as numerous multiple injections cumulating the same dose according to schedules differing in their injection frequencies. Independently of the schedule, the various split-dose regimens spread over weeks/months were as effective as the single delivery of the total dose for mobilization of lung plutonium, and had a therapeutic advantage for removal of retained hepatic and bone plutonium burdens.
We concluded that the cumulative dose level was a therapeutic variable more important than the distribution of split doses for the success of a repeated treatment regimen on retained tissue Pu. Lastly, a pulmonary administration of clodronate which aims at killing alveolar macrophages, thereby releasing their Pu content, associated with a continuous DTPA infusion regimen suggested that the efficacy of injected DTPA in decorporating lung deposit is limited in our conditions because of its restricted penetration into alveolar macrophages and not because of Pu is as a physico-chemical form not available for chelation.

Results presented here were obtained in the frame of the Orano-CEA collaboration.

References

1. L. Grappin, P. Berard, P. Beau, L. Carbone, C. Courtay, X. Castagnet, JP. Le Goff, MO. Neron, J. Piechowski. Radioprotection **42**, 163 (2007)
2. MH. Bhattacharyya, BD. Breitenstein, H. Métivier, BA. Muggenburg, GN. Stradling, V. Volf. Radiat. Prot. Dosim. **41**, 1 (1992)
3. W. Stevens, FW. Bruenger, DR. Atherton, DS. Buster, G. Howerton. Radiat. Res. **75**, 397 (1978)
4. E. Ansoborlo, B. Amekraz, C. Moulin , V. Moulin, F. Taran, T. Bailly, et al. C. R. Chimie **10**, 1010 (2007)
5. GN. Stradling, SA. Gray, JC. Moody, A. Hodgson, KN. Raymond, PW. Durbin, et al. Int. J. Radiat. Biol. **59**, 1269 (1991)

BIO Web of Conferences **14**, 07001 (2019) https://doi.org/10.1051/bioconf/20191407001
HEIR 2018

Relevance and implementation of patient-specific dosimetry in targeted radionuclide therapy

Manuel Bardiès[1, 2]

[1]Centre de Recherches en Cancérologie de Toulouse, Toulouse, France

[2]UMR 1037, INSERM, Université Toulouse III Paul Sabatier, Toulouse, France

1 Introduction

Targeted Radionuclide Therapy (TRT) also called sometimes Molecular Radiotherapy (MRT) is dealing with the therapeutic applications of nuclear medicine. TRT aims at selectively irradiate targets (usually tumours) by the means of a biologic vector labelled with a radioactive isotope. The specificity of the therapy is achieved by the targeting vector, whereas low range radiation irradiates in the neighbourhood of the emission point, thereby insuring selective irradiation of the target and reduced irradiation of surrounding tissues.
According to the vector, one can define TRT specialities:

- Radioimmunotherapy (RIT) uses monoclonal antibodies directed against tumour specific antigens.
- Peptide Receptor Radionuclide Therapy (PRRT) is using peptides directed against some receptors expressed by the tumour cell (e.g. somatostatine receptors in the case of the recently approved Lutathera).
- Metabolic Radiotherapy is a general term to define all situations where the vector participates to the metabolism of the tumour. The therapy with iodine 131 or thyroid pathologies enters in that category. Bone palliation with bone seeking agents also enters into that category.

Selective Internal Radiotherapy (SIRT) of liver tumours with ^{90}Y microspheres does not strictly speaking belongs to TRT, as radioactive microspheres are medical devices administered via hepatic arteries and not radiopharmaceuticals.
The choice of the isotope has been a long matter of debate – and flourishing research – to identify "the" ideal candidate [1]. The perfect isotope should emit low range radiation, in order to preserve the selectivity achieved by the vector. Beta minus emitters with a maximum range between some mm fraction to 1 cm enter in that category. The possible emission of gamma in a range that allows scintigraphic imaging (around 150 keV) and with a rather low yield to avoid non-selective irradiation is a bonus. In addition, the physical half-life of the emitter should be adapted to the biological half-life of the vector, in order to avoid either the premature decay of the radioactive sources, or the premature elimination of radioactive compound. Some commonly used isotopes for TRT include ^{131}I, ^{90}Y, ^{177}Lu, ^{153}Sm. Alpha emitters have also been proposed, but their short range would normally restrict their use to small and easily (quickly) accessible targets, with the notable exception of ^{223}Ra that received FDA/EMA approval for the treatment of bone metastases.

2 Dosimetry of Targeted Radionuclide Therapy

2.1 Relevance of clinical dosimetry for TRT

TRT is usually administered as a "radioactive chemotherapy", where patients are being given the same amount of drug (activity) in a "one size fits all" approach [2,3]. Activity administration (in Bq, or Bq/kg) is defined after "dose" escalation (phase I) clinical trials that assess the toxicity of the drug. However, since the pharmacokinetics of a radiolabelled compound varies from a patient to another, administering *to all* the activity that induces toxicity to the *most fragile* will *undertreat a large fraction of patients*. This calls for patient-specific administrations of the radiolabelled compound, which can be envisaged when it is possible thanks to gamma emissions to follow the fate of the radiolabelled vector in the patient.

The biological effect of procedures involving ionizing radiation can usually be evaluated in the light of the energy delivered to the tissues. The absorbed dose (energy/unit mass, in J/kg or Gy) is an objective index of the irradiation pattern.

2.2 Dosimetry formalism for TRT

According to the MIRD formalism [4], $\overline{D}_k$, the absorbed dose delivered to a target k from radioactive sources in source regions h can be computed as:

$$\overline{D}_k = \sum_h \tilde{A}_h \times S_{(k \leftarrow h)} \quad (1)$$

Where $\tilde{A}_h$ is the total number of decay (cumulated activity) in the source h during the irradiation process (in Bq.s), $S_{(k \leftarrow h)}$ is the S factor, or S value, i.e. the absorbed dose per unit cumulated activity, in $Gy.Bq^{-1}.s^{-1}$. This simplified equation underlines the prerequisites for absorbed dose calculation determination: knowing the number of radioactive sources, within the patient as a function of time and integrating the time-activity curve for all region sources; implementing radiation transport to assess how emitted energy is deposited in surrounding media.

Clinical dosimetry is a chain of operations, each of which impacts the final result (Figure 1).

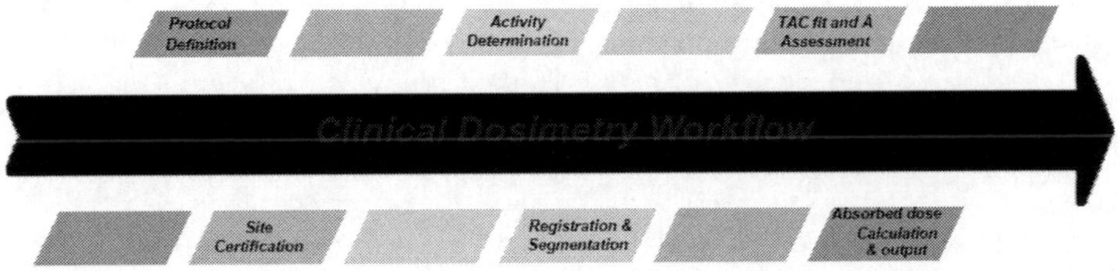

Figure 1: Clinical dosimetry workflow

The relevance of dosimetry in TRT relies on the assumption that an absorbed dose – effect relationship (ADER) can be put in evidence and used to monitor and optimise therapy.

In 2014 the EANM Dosimetry Committee reviewed published ADER evidences [5]. Conclusions were that ADER can be found whenever they are looked for using a relevant/adapted approach. That article also highlighted the fact that absorbed dose calculation is only *one side* of the ADER. The definition of the biological/clinical end-point is also an essential part of the clinical dosimetry protocol.

2.3 Absorbed dose - effect relationships example

For a given clinical endpoint, the way the dosimetry protocol is implemented will impact the conclusion. A very interesting example is given in the context of PRRT in the article of Barone *et al.* [6]. The study considered a subset of patients with neuroendocrine tumours, enrolled in a phase 1 multicentric study intended to establish the maximum tolerated doses of ^{90}Y-DOTATOC in patients with metastatic, somatostatin-receptor positive tumours. That subset of 18 patients benefited from a dosimetric study, in order to evaluate the absorbed doses delivered to the kidneys during a pre-therapeutic administration of ^{86}Y-DOTATOC (3 time-points). The therapy was delivered in 1 to 5 cycles. Therapeutic activity was modulated so that no patient received more than 27 Gy to the kidneys. Kidney toxicity was assessed by monitoring kidney function (serum creatinine and creatinine clearance).
PET quantitative imaging with ^{86}Y was challenging and eventually led to a specific publication[ix].

- Absorbed doses were initially computed using MIRDOSE 3.1 [7], under the assumption that all patients had the same kidney volume (but differentiating between male and female patients).
- A second (retrospective) study considered the actual volume of each patient's kidney, and rescaled the S values obtained from MIRD pamphlet 19 [8] to a patient-specific S value (and considered the absorbed dose delivered to the cortex rather than that to the whole kidney).
- A third approach included radiobiological parameters (from EBRT) to derive the Biologic Effective Dose (BED) delivered to the kidney cortex.

The study to the correlation between the loss of kidney function (toxicity) and either the kidney absorbed dose or BED was presented Barone *et al*. It clearly demonstrates how the dosimetric protocol impacts ADER.

In addition, from a dosimetric point of view, it also highlights how ALL aspects that lead from activity quantification to absorbed dose (or BED) matter to put in evidence ADERs: even a very refined activity determination (based on quantitative ^{86}Y-DOTATOC PET imaging) will not provide good ADER if the same S value is used for all patients...

3 Conclusions

Further analysis of published evidence shows that there is no and *there won't be a generic clinical dosimetry procedure*. "Dosimetry that works", i.e. providing for ADERs is disease/application/radiopharmaceutical specific.

The fact that there is no reference clinical dosimetry protocol generic enough to be adapted to all therapeutic applications of nuclear medicine should not prevent us from implementing what seems relevant at a given moment, using available resources. Documenting the protocol is an essential part of the reporting and should be taken as seriously as possible [9].

References

[1] Mausner LF, and Srivastava SC. Selection of radionuclides for radioimmunotherapy. Medical Physics. 1993;20(2).

[2] Chiesa C, Sjogreen Gleisner K, Flux G, Gear J, Walrand S, Bacher K, Eberlein U, Visser EP, Chouin N, Ljungberg M, Bardiès M, Lassmann M, Strigari L, and Konijnenberg MW. The conflict between treatment optimization and registration of radiopharmaceuticals with fixed activity posology in oncological nuclear medicine therapy. Eur J Nucl Med Mol Imaging. 2017;44(11): 1783-1786.

[3] Flux G, Sjogreen Gleisner K, Chiesa C, Lassmann M, Chouin N, Gear J, Bardiès M, Walrand S, Bacher K, Eberlein U, Ljungberg M, Strigari L, Visser E, Konijnenberg M. From fixed activities to personalized treatments in radionuclide therapy: lost in translation? Eur J Nucl Med Mol Imaging. 2018; 45(1):152-154.

[4] Loevinger R, Budinger TF, Watson EE. MIRD Primer for Absorbed Dose Calculations. Revised ed. New York, NY: The Society of Nuclear Medicine; 1991.

[5] Strigari L, Konijnenberg M, Chiesa C, Bardiès M, Du Y, Gleisner KS, Lassmann M, and Flux G. The evidence base for the use of internal dosimetry in the clinical practice of molecular radiotherapy. Eur J Nucl Med Mol Imaging. 2014;41(10):1976-88.

[6] Barone R, Borson-Chazot F, Valkema R, Walrand S, Chauvin F, Gogou L, Kvols LK, Krenning EP, Jamar F, and Pauwels S. Patient-specific dosimetry in predicting renal toxicity with 90Y-DOTATOC: relevance of kidney volume and dose rate in finding a dose-effect relationship. Journal of Nuclear Medicine. 2005;46(1 suppl):99S.

[7] Stabin MG. MIRDOSE: personal computer software for internal dose assessment in nuclear medicine. J Nucl Med. 1996;37:538 -546.

[8] Bouchet LG, Bolch WE, Blanco HP, Wessels BW, Siegel JA, Rajon DA, Clairand I, and Sgouros G. MIRD Pamphlet No 19: absorbed fractions and radionuclide S values for six age-dependent multiregion models of the kidney. J Nucl Med. 2003;44(7):1113-47.

[9] Lassmann M, Chiesa C, Flux G, Bardiès M, and EANM Dosimetry Committee. EANM Dosimetry Committee guidance document: good practice of clinical dosimetry reporting. Eur J Nucl Med Mol Imaging. 2011;38(1):192-200.

^{90}Y glass microspheres radionuclide therapy: robustness analysis with a dosimetry software

Aurélie Forbes[1,*], *Gilles* Grimon[1], *Aya* Esaki[1,2], *Jad* Farah[1], and *Emmanuel* Durand[1]

[1]APHP Bicêtre University Hospital, Biophysics and Nuclear Medicine Department, 78 rue du Général Leclerc, 94275 Le Kremlin-Bicêtre Cedex, France
[2]CentraleSupélec, 3 Rue Joliot Curie, 91190 Gif-sur-Yvette, France

1 Introduction

In diagnostic and therapeutic nuclear medicine, the administration of a fixed radionuclide activity simply adjusted to patient's weight remains a widely used practice. To perform personalized dosimetry, several home-made solutions exist [1] while new commercial solutions have been recently introduced with treatment planning and dosimetry capabilities similar to external radiotherapy.

In the present work, the PLANET® Dose (DOSIsoft, France) software was used to retrospectively compute organ doses for eight liver cancer patients treated at Bicêtre University Hospital using glass ^{90}Y Theraspheres™ (BTG, Canada). The standard protocol for radioembolization involves a planning phase performed with ^{99m}Tc-macroaggregated albumin (MAA). SPECT/CT images are acquired to determine the liver tumor volume and the liver-to-lungs shunt [2]. The ^{90}Y activity is then calculated based on a simple 2-compartment model (lungs and targeted liver regions) with a planned mean absorbed dose of 80 - 150 Gy (120 Gy recommended by BTG) to the target and a maximum tolerable dose to the lungs of 30 Gy. Following ^{90}Y injection, Theraspheres™ distribution is verified through PET/CT images while no organ dose calculation whatsoever is performed. This study aims at computing and comparing planned (standard approach) and delivered organ doses (PLANET® Dose).

2 Materials and methods

Using PLANET® Dose dosimetry software (DOSIsoft, France), a proper segmentation of liver and lungs organs was performed for eight patients treated between February 2016 and may 2017 using ^{90}Y Theraspheres™. From the registered SPECT (functional) and CT (anatomic) planning images, tumor (TV) and non-tumor (NTV) volumes and lungs structures were identified and their volume was computed. Liver-to-lungs shunt was also computed using PLANET® Dose and compared against the standard procedure. Organ doses and Dose Volume Histograms (DVH) were next calculated using the convolution or local deposition methods. Such calculations were performed both on predictive MAA SPECT/CT acquisitions as well as on post treatment ^{90}Y PET/CT images to determine the correspondence between planned and delivered tumour and lungs doses.

* Corresponding author: aurelie.forbes@aphp.fr

3 Results

Results first show that the liver-to-lungs shunt based on planar SPECT images with low MAA lungs fixation is generally higher (on average 40 ± 10%) than the one calculated with PLANET® Dose using volumetric information. This might requires revising the standard approach since proper assessment of liver-to-lungs shunt would enable TV dose escalation. However, in the specific eight patients considered in this study, the impact on the activity administration and the standard clinical procedure remains limited since the mean liver-to-lungs shunt was low (4.2 ± 2%). In addition, some patients SPECT-CT images used with PLANET® Dose had missing upper part of lungs due to local patient positioning.

The distributions of ^{99m}Tc MAA and ^{90}Y Theraspheres™ agreed in 5/8 patients. When the distributions disagree, PLANET® Onco provides a quantitative value of the delivered TV dose; this information is strictly missing in the standard approach. Namely, predictive dosimetry was found giving a higher mean TV dose by a factor 1.44 ± 0.34; a similar finding of Gnesin et al. [3]. Such under-dosage to TV could yield a procedure failure.

Finally, convolution and local deposition calculation methods showed similar (average difference 2 ± 2%) doses to tumor and non-tumor further justifying the performance of home-made solutions which often use either algorithms.

4 Conclusion

In ^{90}Y radioembolization, the dose delivered to target and non-target organs is not quantitatively calculated with conventional treatment planning. Moreover, the predictive distribution of ^{90}Y Theraspheres™ is sometimes in disagreement with ^{99m}Tc MAA. PLANET® Dose software enables personalized dosimetry and DVH calculations.

Following this study, a systematic organ dose calculation based on ^{90}Y Theraspheres™ PET/CT images proved necessary to advise on the appropriate actions: adjunct treatment in case of TV under-dosage or follow-up for healthy organs' over-exposure.

The robustness analysis presented in this preliminary study will be integrated into the routine local practice to adjust, in future treatments, both the planning and post treatment patient follow-up.

Further studies remain to be done to assess whether this software gives better clinical evaluation than the standard method.

We acknowledge DOSIsoft for their support in using PLANET® Dose software.

References

1. C. Chiesa et al. Radioembolization of hepatocarcinoma with 90-Y glass microspheres: development of an individualized treatment planning strategy based on dosimetry and radiobiology. Eur J Nucl Med Mol Im. 2015;42:1718–38.
2. BTG, TheraSphere™ Yttrium-90 Glass Microspheres, Instructions for Use.
3. S. Gnesin et al. Partition Model-Based 99mTc-MAA SPECT/CT Predictive Dosimetry Compared with 90Y TOF PET/CT Posttreatment Dosimetry in Radioembolization of Hepatocellular Carcinoma: A Quantitative Agreement Comparison. J Nucl Med. 2016;57(11):1672–8.

Radio-iodine therapy of Graves' disease: is there a dose-effect relationship?

David Broggio[1,1]

[1]IRSN, IRSN/SDOS/LEDI, BP-17, 92626 Fontenay-aux-Roses, France

1 Introduction

Graves' disease is a benign disease very often treated with radioiodine [1], its incidence is about 30/100 000/year in Europe and USA [2]. Even if several millions of radioiodine treatments have been administrated [1] there are still debates on the exact amount of iodine that should be administrated and on the final goal of the therapy: permanently hypothyroid or euthyroid [3-4].

Inspection of the literature shows that there are little evidences for choosing the requested thyroid dose. In order to obtain the dose-effect relationship for the success of iodine-131 therapy for Grave's disease we carried out an analysis of the literature.

2 Materials and Methods

We looked for literature data reporting the success of a single I-131 therapy for Grave's disease. We only included peer-reviewed articles in French or German where the thyroid dose was reported along with the success rate.
We assigned an uncertainty to the success rate (exact Cloper-Pearson method) and to the reported dose. We finally carried out a fit of the data using, as an ansatz, a sigmoid function. In the fitting procedure both error bars were taken into account following the method of Wolberg [5].

3 Results and discussion

We extracted from the literature 22 relevant papers accounting for a total of 3922 patients. The reported thyroid doses were between 60 and 649 Gy. In all studies the thyroid dose was calculated with at least one uptake measurement and an ultrasound assessment of the thyroid weight. In 15 of the studies the dose uncertainty was reported or could be deduced. An average dose uncertainty of 26% was fixed for the other studies.

The resulting fitted dose-effect relationship is shown in figure 1. The data supporting the fit are relatively scattered. However, the fitting procedure naturally produces and average trend and the associated 95% confidence band. The fitted result is in very good agreement with the 1995 data of Peters [7] which were obtained with 98 patients (data not shown).

---

[1] Corresponding author: david.broggio@irsn.fr

According to our findings the confidence bands for therapy success are as follows: [55-80%] at 150Gy, [65-90%] at 200Gy and [80-95%] at 300Gy.

Several factors can explain the scattered pattern of the obtained data: large uncertainties on axes, different follow-up periods, different regimen of medication withdrawal before therapy. Surprisingly, such a scattered pattern can also be found in multi-centre studies in external radiotherapy.

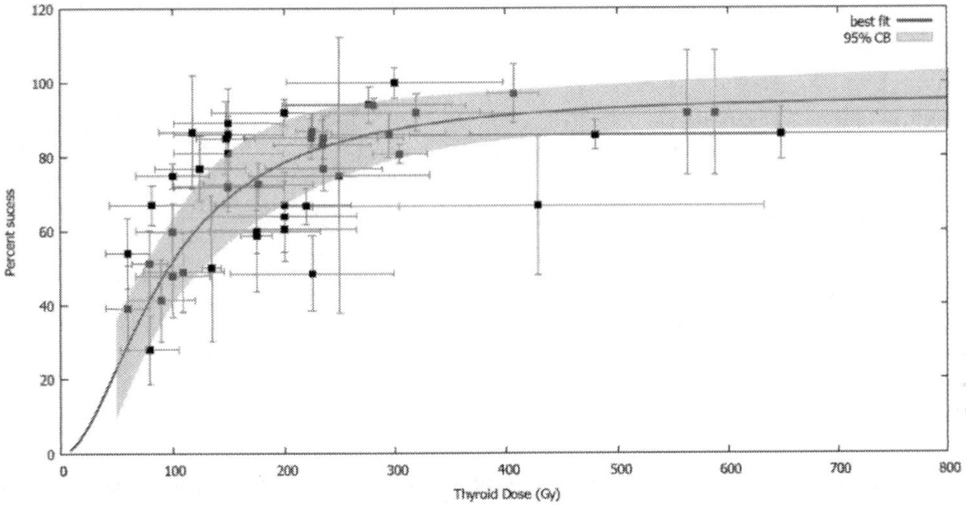

Figure 1. Fitted dose-effect relationship: success rate as a function of thyroid dose.

4 Conclusion

To the best of our knowledge this work is the more exhaustive and complete analysis of the dose-effect relationship for one of the most common radio-iodine therapy. On one hand, if one assumes that the data and analysis are reliable it should be concluded that there is no real dose-effect relationship. One the other hand, assuming that a dose effect relationship must exist, it should be concluded that in future studies the dose assessment should be more accurate, using for example several uptake measurements, and that follow-up times and inclusion criteria chosen more rigorously.

References

1. Bonnema S.J and Hegedüs L. Endocrine Reviews, 33 (6): 920-980 (2012).
2. McLeod D.S.A. and Cooper D.S. Endocrine **42** (2): 252-265 (2012)
3. Van Isselt J.W *et al.* European Journal of Nuclear Medicine and Molecular Imaging **34** (11):1883-1884 (2007)
4. Kalinyak J.E. and McDougall I.R. Journal of Clinical Endocrinology and Metabolism **88**(3):975-977 (2003).
5. Wolberg J. Data Analysis Using the Method of Least Squares, 2006, Springer
6. Peters H. et al. European Journal of Clinical Investigation **25** (3): 186-193 (1995).

Absorbed dose evaluation at different organs after ^{131}I oral contamination of two Wistar rat models

Meriem Mezaguer-Lekouaghet[1*], *Eric* Blanchardon[2], *Abdelwahab* Badreddine[1], *Jean-Marc* Bertho[2], *Maamar* Souidi[2], *Ahcène* Baz[3], and *Zohra* Lounis-Mokrani[1]

[1] Centre de Recherche Nucléaire d'Alger, Algérie
[2]Institut de Radioprotection et de Sûreté Nucléaire Fontenay-Aux-Roses, France
[3]Ecole Normale Supérieure de Kouba, Algérie

*Corresponding author: m_mezaguer@yahoo.fr

Abstract

Iodine-131 (^{131}I) is one of the most frequently used radionuclides for diagnosis and therapy of thyroid diseases. It is administered orally in the treatment of cancer to eliminate the residual postoperative microscopic tumor foci, and the residual normal thyroid tissue for early detection of recurrence [1].

The comparative behavior of ^{131}I concentration into two animalmodels with total and partial thyroid has been investigated in our previous work [2]. The accumulated activities have been measured in fourteen organs.

In this study, the mean absorbed doses resulting from ^{131}I accumulated in all organs have been evaluated using RODES software [3, 4]. With this software, mean absorbed doses were calculatedfor selected organs (thyroid, lungs, heart, liver, kidneys, stomach, spleen, large and small intestine, testes, urinary bladder wall) by combining the specific absorbed fractions (SAF) of energy with radiation emission spectra and biokinetic data determined from our previous experimental study [2]. Calculations were based on the ^{131}I photon and electron emissions reported by [5] and SAFs previously calculated by Monte Carlo simulation in the voxel phantom of an adult male rate [3, 4].

The obtained results show high absorbed dosesdeliveredto stomach and lungs for both models compared to other organs. The dose received by the testes and salivary glands is found to be higher in the case of the rat model without thyroid. Conversely, the spleen and bladder wall received lower doses in this latter model compared to those received by the rat model with thyroid. One can also note that the difference in mean absorbed dose received by liver, lungs, heart, and walls of the stomach is not significant between the two rat models.

References

[1] S.P. Marcel, D.H. Junak, M. Lassmann, M. Dietlein, M. Luster, *Eur J Nucl Med Mol Imaging*, 37(2010).

[2] M. Mezaguer-Lekouaghet, A. Badreddine, N. Ait Kaci, K. Remil, A. Benbournane, F. Boucheneb, F. Messili, A. Baz, M. Souidi, & Z. Lounis-Mokrani, *ICNTRM. Strasbourg France* (2017).

[3] M. Locatelli, H. Miloudi, G. Autret, D. Balvay, A. Desbrée, E. Blanchardon, and J.M Bertho, *J. Radiol. Prot.* 37 (1) (2017).

[4] H. Miloudi, M. Locatelli, G. Autret, D. Balvay, A. Desbrée, E. Blanchardon, and J.M Bertho *J. Radiol. Prot.* 37 (3) (2017).

[5] International Commission on Radiological Protection. *ICRP Pub.107.* (2008).

BIO Web of Conferences **14**, 07005 (2019) https://doi.org/10.1051/bioconf/20191407005
HEIR 2018

Ruthenium-106 brachytherapy in treatment of uveal melanoma: 50 years of use!

Sarah Le Hir[1], and *Florence* Menetrier[1]

[1]CEA/DRF/D3P/Prositon, 18 Route du panorama, 92265 Fontenay-aux-Roses Cedex, France

Ruthenium-106 (^{106}Ru) is a beta emitter with a 374 day half-life that is commonly used for brachytherapy of uveal melanoma. Since the beginning of its use in the 1960s, it has been the treatment of choice in Europe unlike North America where it is disfavoured. Compared to other isotopes including beta (Sr-90) and gamma (I-125, Pd-103) emitters, it has a limited penetration depth but the advantage of sparing surrounding normal tissues. Ru-106 plaque radiotherapy has been increasingly used for the treatment of small- to medium-sized tumors up to 5 mm in thickness [1]. ^{106}Ru plaque provides a maximum activity of 1.6 mCi and a recommended radiation dose to the apex of tumors is 85 Gy according to a Collaborative Ocular Melanoma Study (COMS) [2].

The local control, non-recurrence and eye-conserving are the main concerns in uveal tumor treatment, besides the patient survival. Though, the radiation threshold dose, effective for the treatment remains unknown and further examinations are needed to design an optimum dose radiation.

Taking into account of geometric tumour parameters, the dose and the time of treatment are directly correlated to the height and basal diameter of tumors, respectively. Software calculations are designed and developed to provide a robust and optimal dose regarding a loss of visual acuity. In fact, dosimetric tolerance levels to macula, optic nerve and retina are evaluated. The Monte Carlo algorithms are used to simulate a dose distribution in the tissues. Indeed, they need to simulate a spectrum resulting from a beta disintegration complicated by a heterogeneous source distribution. Each device used is beforehand calibrated and dose rate in water phantom is given.

Nevertheless, an assessment of dose calculations into plaques points a difference between homogeneous and heterogeneous plaques. A dose distribution of ^{106}Ru plaque and the normalized surface distribution of the emitter substance into a heterogeneous plaque are illustrated on Fig. 1.

A state of art of the use of this radionuclide in terms of dose/response and risk assessment will be presented and the results will be discussed.

https://doi.org/10.1051/bioconf/20191407005

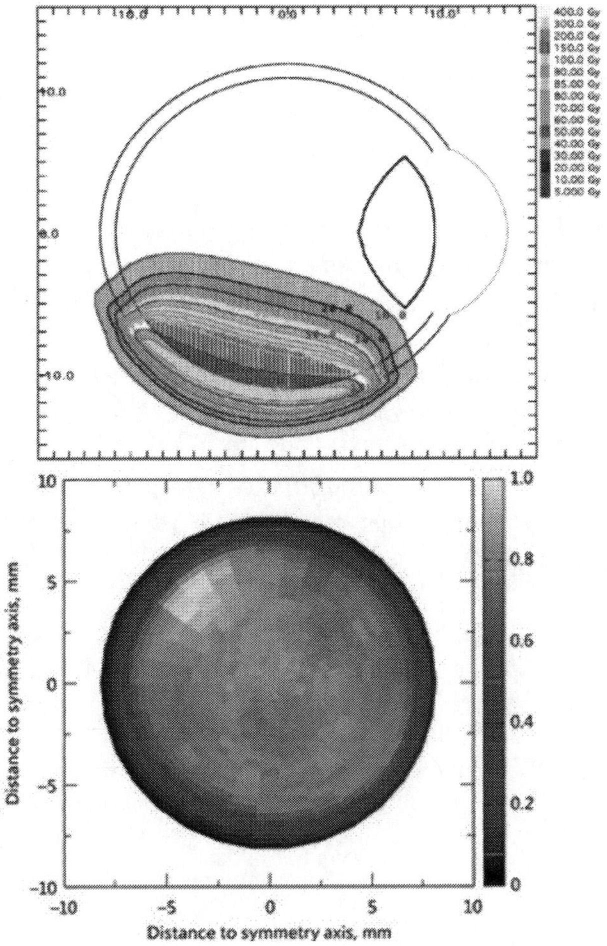

Fig. 1: 2D dose (Gy) distribution for 18 mm ^{106}Ru plaques [3] (left), experimental emitter dose map for heterogeneous plaque (CCA1364) [4].

References

1. Stannard, C., et al., *Radiotherapy for ocular tumours.* Eye, **27**(2): p. 119-127 (2013)
2. Melia, B.M., et al., *Collaborative ocular melanoma study (COMS) randomized trial of I-125 brachytherapy for medium choroidal melanoma. I. Visual acuity after 3 years COMS report no. 16.* Ophthalmology,. **108**(2): p. 348-66 (2001)
3. Wilkinson, D.A., et al., *Dosimetric comparison of Ru-106 and I-125 plaques for treatment of shallow (<= 5 mm) choroidal melanoma lesions.* British Journal of Radiology, **81**(970): p. 784-789 (2008)
4. Zaragoza, F.J., et al., *Monte Carlo Estimation of Absorbed Dose Distributions Obtained from Heterogeneous (106)Ru Eye Plaques.* Ocul Oncol Pathol, **3**(3): p. 204-209 (2017)

Comments on actinide radiotoxicology research and the 3Rs remit – Replace, Reduce and Refine

Nina M. Griffiths[1,*], *Anne* Van der Meeren[1], *Jaime F.* Angulo[1]

[1]Laboratoire de RadioToxicologie, CEA, Université Paris-Saclay, 91297 Arpajon, France

1 Introduction

Radiotoxicology addresses the diverse biological effects and handling of radionuclides in humans, animals and in the environment. To characterise the health effects of incorporated radionuclides (HEIR) many experiments have been conducted in different animal models. Alpha particle-emitting actinides, such as plutonium, have been (and still are) the focus of many experiments with diverse aims and in multiple species. These range from (1) applied (quantification of tissue uptake/retention, decorporation therapy efficacy), (2) fundamental (gene expression, cancer induction), (3) biokinetic models for dose estimation and (4) environmental (food chain accumulation). The ICRP have used animal data, in part, for development of different biokinetic models to improve dose calculations in humans. Current societal demands now question the use of *in vivo* approaches for risk assessment, an integral part of actinide radiotoxicology research. Future research is needed concerning biokinetics and decorporation strategies of unknown, complex radioelements from different environments of new nuclear fuel and decommissioning. It is now timely to incorporate the replacement, refinement and reduction of animals (the 3Rs) for this research.

2 The 3Rs

The concept of the 3Rs, developed by Russell and Burch, has become embedded in the guidelines and legislation for animal experimentation in many countries. This concept encompasses *Replacement* – use of alternative methods (cell culture, mathematical modelling …), *Reduction* – use of an adequate number of animals to generate significant informative and reproducible data – requires good experimental design and help from statisticians, *Refinement* – use of improved animal welfare and husbandry, reduced invasiveness and stress as well as better assessment of pain. This presentation has for objective to privilege the remit of the 3Rs in radiotoxicology. The aim however is not to advocate prohibition of animal experiments, no model has the perfect answer but to present examples of specific questions pertaining to actinide radiotoxicology.

3 Actinide biokinetics, biological effects and decorporation

Biokinetics Actinide biokinetics and transfer from exterior to interior across the pulmonary barrier are governed by the physicochemical nature of the compound. Extensive

* Corresponding author: nina.griffiths@cea.fr

in vivo experiments have been carried out using pulmonary actinide administration to evaluate actinides ranging from insoluble ("S") to soluble ("F") forms. What answers could be given by replacement *in vitro* approaches using different actinide physicochemical forms? Simple approaches using dissolution techniques provide invaluable information on solubility and transferability. However, more complex *in vitro* models are required to mimic better cellular and acellular environments that may interact with actinides. For example, part of the soluble fraction may be bound to lung parenchyma "lung bound fraction" and further *in vitro* experiments can be used to determine respective affinities for biological ligands. Alveolar macrophages play a major role in lung actinide biokinetics to take up insoluble particles and again *in vitro* experiments can provide information on the differential uptake of various actinide physicochemical forms. In animal experiments the initial amount of actinide is known, the form is known and excretion is measured as are actinide tissue levels – all of have provided invaluable information for model construction. Historic data from humans (USTUR) or animal experiments are continually reassessed due to improved mathematical analyses and as such, actinide radiotoxicological research has been following at least two elements (replace, reduce) of the 3Rs remit for many years.

Biological effects Inhalation of actinides is a potential health risk for people handling actinides for either civil or military purposes. The different biological effects, particularly lung tumour induction, of these alpha emitters have been extensively studied in many different animal species since initial production of plutonium. The biological effects of retained actinides, either in pulmonary epithelial cells or in macrophages may be evaluated *in vitro* and compared to historic experimental or human data (e.g. actinide-induced lung tumours). Advances in cell culture, particularly 3D-printed cell culture platforms as well as stem cells can perhaps answer pertinent questions on lung dosimetry, lung binding and lung effects of actinides. These approaches could facilitate real time analyses without using large numbers of animals as it is always important to consider that lung tumours are an end-point and the questions of initiating events and which cells are irradiated by the alpha particle-emitting actinides remain unresolved. This is the concept of Adverse Outcome Pathways.

Decorporation Many animal experiments have also been carried out to evaluate the efficacy of potential decorporation agents. DTPA is the recognised treatment for decorporation of certain actinides (Pu, Am, Cm). Other approaches are feasible and have indeed been used – chemical affinity, molecular design and *in vitro* studies. However, potential new compounds may be very good chelators in a test tube in non-physiological media, but may be sequestered rather than eliminated, may produce toxic metabolites or may form an unstable actinide-chelator complexes *in vivo*. In this case, animal experiments can provide important pharmacokinetic and pharmacodynamic data. Moreover, the question of whether it is more advantageous to administer DTPA to the lungs directly compared to systemic dosing is important. Again judicious planning of animal experiments may allow the elaboration of a better biokinetic model for actinide behaviour associated with DTPA.

4 Conclusions

Within the 3Rs remit future research on actinide radiotoxicology should incorporate (1) Replace – development and use of appropriate, validated *in vitro* models where possible, (2) Reduce – use of *in vitro*, *ex vivo* and *in silico* models for "predictive screening" of potential decorporation or decontamination agents before *in vivo* testing and (3) Refine – use of improved experimental procedures, animal welfare with a prior harm-benefit analysis. Future research on actinides *in vivo* taking into consideration the 3Rs remit will undoubtedly be necessary to address the question of complex actinide forms likely to be encountered in new nuclear fuels and nuclear facility decommissioning.

FAIRing the radiation science commons

Balázs Madas[1,1], *Ulrike* Kulka[2], *Bernd* Grosche[2], *Mandy* Birschwilks[2], *Gayle* Woloschak[3], *Shin* Saigusa[4], *Soile* Tapio[5], *Michael* Gruenberger[6], *Paul* Schofield[6]

[1]Environmental Physics Department, MTA Centre for Energy Research
H-1121 Budapest, Konkoly-Thege Miklós út 29-33.
[2]Department Radiation Protection and Health Federal Office for Radiation Protection (BfS)
Ingolstaedter Landstrasse 1 85764 Oberschleissheim Germany
[3]Departments of Radiation Oncology, Radiology, and Cell and Molecular Biology, Robert E. Lurie Comprehensive Cancer Center, Feinberg School of Medicine, Northwestern University 303 E. Chicago Ave., Ward 13-002, Chicago, IL 60611
[4]Institute of Radiation Epidemiology, Radiation Effects Association, 1-9-16 Maruishi Dai-ni Building 5F, Kajicho, Chiyoda-ku, Tokyo, 101-0044, Japan
[5]Helmholtz Zentrum München , Deutsches Forschungszentrum für Gesundheit und Umwelt (GmbH) , Ingolstädter Landstr. 1 , 85764 Neuherberg , Germany
[6]Dept of Physiology Development and Neuroscience, University of Cambridge, Downing Street, Cambridge CB2 3EG, UK

The FAIR principles

The primary data produced in the course of publicly-funded science represents a common asset for society as much as the analysed and interpreted results. Recent years have seen a unanimous agreement that such data and discoveries should be as accessible as possible by other scientists and the members of society in order to extract the maximum value from that investment. The concept of the science commons is well established and legal economic and social aspects of the commons are the subjects of intensive interest and examination [1]. Open access publication has been the focus of much attention and most investigators are now familiar with mandated open access publication mechanisms and requirements. However the public availability of primary data particularly requires the establishment of rules of governance and a sharing infrastructure. Through sharing the value of research commons are enhanced through the data cycle with iterated use and recontribution of new or modified data and materials creating an amplification effect that enhances the value and ultimately sustainability of the research commons.

The governance and infrastructure needed to support a research commons are two critical arms of successful scientific enterprise. The FAIR guidelines for data sharing were established in 2015 as a consequence of extensive consultation and apply to the community through consent and consensus [2]. The basic rules of FAIR are articulated in 15 areas summarised under the four letters of FAIR -Findable, Accessible, Interoperable, Reusable. The implications of the FAIR guidelines are that data should be discoverable and accessible by a human or by machine, that it should have sufficient metadata to be understandable and implementable and critically that the originator of the data should not be involved in the

[1] Corresponding author: balazs.madas@energia.mta.hu

decision as to whom it is made available. The FAIR principles do not preclude licensing or reasonable charges for access, so FAIR does not necessarily mean free or free of constraint over use, but that data should be accessible under reasonable conditions and in fact most of the data that concerns us lies within the pre-competitive space in any case. Major funding agencies such as the Wellcome Trust and the European Commission [3], and the NIH are now also trialing a FAIR data commons policy [4].

Data sharing in radiation science

Databases and repositories are the essential infrastructure for the research commons and require coordinated development and sustainable funding [5, 6]. In 2009 the STORE database was first set up under European Commission funding to encourage public data sharing and reuse in the domain of radiation biology. It was developed through successive grants and was opened to public use in 2014. STORE provides a sharing infrastructure for all kinds of data types and domains, ranging from epidemiology and human cohort data to 'omics and cytogenetics. Increasingly STORE is being used by large distributed projects to coordinate and archive primary and derivative data which is then used for support of publications. STORE mints persistent digital object identifiers and accession IDs which use a namespace formally registered with identifiers.org at the EBI. Similarly registered with the FAIR sharing initiative and r3data [7] STORE is a well recognised and accepted data repository. The database is physically located at the Bundesamt fuer Strahlenschutz in Munich and has the full security of a German Federal core data service. The BfS has undertaken to maintain the database indefinitely which means that data will be secure and accessible for the foreseeable future. Currently STORE contains around 3000 data objects across a wide range of data types; the number are increasing rapidly. The aim of STORE is to promote open access and reuse of data, as well as the archiving of at-risk or legacy data, thus promoting and enhancing the scientific commons. Consequently deposition and access to data are free to individual investigators and to funding agencies. STORE is available on http://www.storedb.org and users may sign up using their ORCID ID.

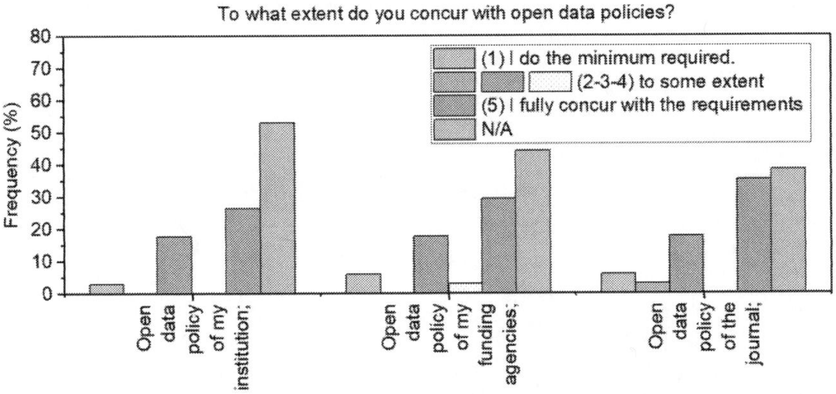

Figure 1. Concurrence of the respondent to open data policies.

Our experience of working with the radiation science community during the development of STORE has raised questions concerning the level of training in research data management, the culture of data sharing and various perceived disincentives for sharing [8]. This conflict with the development of policies of funding agencies and, increasingly, journals is a source of significant concern and we consider recommendations to address this problem through training, particularly of early-career scientists.

References

1. A. Mishra and T. Bubela, *OMICS: A Journal of Integrative Biology,* **18**, 254 (2014).
2 M. D. Wilkinson, M. Dumontier, I. J. Aalbersberg, G. Appleton, M. Axton, A. Baak, *et al., Sci Data,* **3**, 160018, (2016).
3. H2020 Guidelines on FAIR Data Management:

http://ec.europa.eu/research/participants/data/ref/h2020/grants_manual/hi/oa_pilot/h2020-hi-oa-pilot-guide_en.pdf
4. https://commonfund.nih.gov/commons
5. P. N. Schofield, J. Eppig, E. Huala, M. H. de Angelis, M. Harvey, D. Davidson, *et al.*, *Science,* 330, 592, (2010).
6. S. A. Sansone, P. Cruse, and M. Thorley, *Sci Data,* **5**, 180027, (2018).
7. https://www.re3data.org/about
8. B. G Madas, P. N. Schofield, Survey on data management in radiation protection research arXiv:1805.05463 [cs.CY] Accepted for publication in *Radiat Prot Dosim*.

USTUR: Expanding horizons for actinide biokinetics and dosimetry

Sergei Y. Tolmachev[1, *], *Stacey L.* McComish[1], and *Maia* Avtandilashvili[1]

[1] U.S. Transuranium and Uranium Registries, College of Pharmacy, Washington State University, 1845 Terminal Drive, Suite 201, Richland 99354, WA, USA

Abstract

Since 1968, the U.S. Transuranium and Uranium Registries (USTUR) has followed up with occupationally-exposed individuals (volunteer Registrants) by studying the biokinetics (deposition, translocation, retention, and excretion) and tissue dosimetry of actinide elements [1].

The USTUR holds data on work history, radiation exposure and bioassay measurements, as well as medical records from more than 400 former nuclear workers. These individuals had documented intakes of actinides at the levels higher than 74 Bq. Inhalation and wound are two major routes of intake and ^{239}Pu is a primary radionuclide (Fig. 1.)

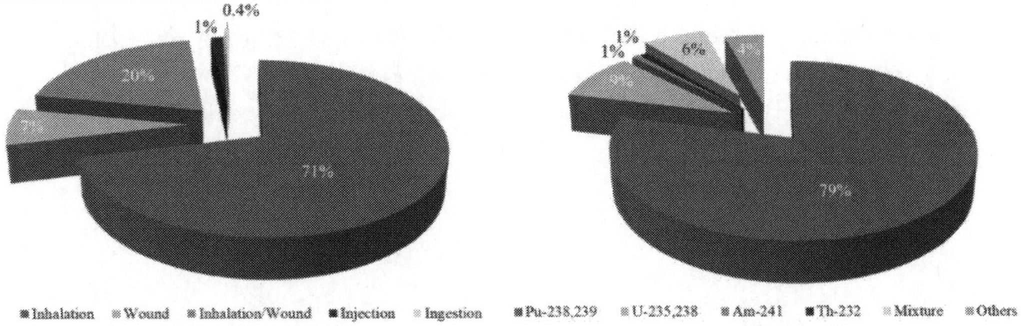

Fig. 1. USTUR Registrants' exposure by: route of intake (left) and primary radionuclide (right).

Post-mortem radiochemical analyses of tissues obtained at autopsy, especially those from the whole-body donors, allows USTUR to significantly improve our knowledge of distribution and long-term retention of actinides in the human body and have helped in parameterizing biokinetic constants for these radioactive elements.

Recently, several groups of individual cases have been identified for studying biokinetics and dosimetry for specific radionuclides, exposure scenarios and materials, as well as effects of decorporation treatment (Table 1).

* Corresponding author: stolmachev@wsu.edu

Table 1. USTUR specific study groups.

Study group	Number of cases
^{239}Pu	
Soluble	14
Refractory	22
^{238}Pu	10
^{241}Am	3
Uranium	
Enriched	5
Depleted	3
Natural	17
^{232}Th	3
^{237}Np	1
^{244}Cm	1
^{239}Pu wound	14
Decorporation	
^{239}Pu	14
^{238}Pu	1
^{241}Am	2

Data from a USTUR whole-body donor who was exposed to soluble ^{239}Pu via inhalation were used to study long-term plutonium retention in the upper airways and quantify the plutonium 'bound' fraction [2, 3]. This is only one notable example of the manner in which USTUR data can be used to improve accuracy of dose assessment and radiation protection of plutonium workers.

References

1. The United States Transuranium and Uranium Registries. https://ustur.wsu.edu/
2. A. Birchall, M. Puncher, S. Y. Tolmachev. Radiat. Prot. Dosim. **176**, 50 (2017).
3. A. Birchall, M. Puncher, A. Hodgson, S. Y. Tolmachev. Health Phys. Published Ahead-of-Print (2018).

Internal exposure in post-accidental situations: a multi-pollution to take into account?

Jean-Marc Bertho[1,*], *Stefania* Musilli[1], and *Karine* Tack[1]

[1] IRSN, PSE-SAN/SESANE, LRTOX, BP 17, 92262 Fontenay aux roses cedex, France

1 Introduction

Major nuclear accidents (Chernobyl, Fukushima) induced the contamination of large areas with a mix of radionuclides (RNs). On the long term the exposure of the population is limited to some key RNs such as ^{137}Cs, ^{134}Cs and ^{90}Sr and to external exposure at low dose and low dose rate.

Numerous studies, either epidemiological or more limited, were conducted on the health effects of long term chronic ingestion of RNs. Epidemiological studies showed the occurrence of leukemia both in liquidators and in the general population. Other somatic pathologies were also evidenced such as cardiovascular diseases and cataracts. More limited studies showed the occurrence of other somatic disorders such as changes in the immune system in human population living on these contaminated territories. In numerous cases, the observed effects were ascribed to the internal contamination with ^{137}Cs. However, the direct link with ^{137}Cs internal exposure or with external exposure was not always clearly established. Moreover, in most of epidemiological studies, the assessment of exposure in post-accidental situation remains difficult with numerous uncertainties. As a result, the health effects of incorporated radionuclides by populations living in a contaminated territory remain a matter of controversy.

2 Experimental studies in animal models.

In order to overcome the uncertainties in exposure assessment, several groups studied the effect of chronic ingestion of radionuclides in rodents, with various schedule of exposure, either acute or chronic. Since ^{137}Cs is the most frequent RN found in the environment, this RN was the most studied. After an acute exposure, it was shown the appearance of an hematopoietic syndrome in a dose dependant manner. As well, the frequency of tumors was increased according to the ^{137}Cs dosage. Studies of chronic exposures through ingestion showed the appearance of numerous biological effects, both in the rat and in the mouse models. However most of these effects were of limited amplitude and did not led to the appearance of pathologies. This was the case for instance for the oxysterol metabolism with changes evidenced in steroid hormone synthesis, Vitamin D metabolism and hepatic cholesterol metabolism, and for the sleep wake cycle. The main change observed was a change in the cardiovascular system with a decrease in the arterial pressure and a disappearance of the circadian rhythm of this AP.

* Corresponding author: jean-marc.bertho@irsn.fr

These experimental results are somewhat in contradiction with results from the study of human populations. In fact, there is no evidence of a cardiovascular dysfunction induced by exposure to ^{137}Cs in the general population. As well, there are numerous studies in humans exposed to ^{137}Cs showing alterations in the immune system while in the rodent model, no change was evidenced.

A possible hypothesis to explain these discrepancies is the uncertainties in the exposure of populations in a post-accidental exposure. For instance, ^{90}Sr, which is also present in the environment of territories contaminated by the Chernobyl accident, was recently shown to induce changes both in bone physiology and in the immune response induced by an antigen in a vaccine model in rodents. These results are in accordance with results of large studies in the population living on the riverside of the Techa River, heavily contaminated with ^{90}Sr, among other radionuclides. Of course, these results demonstrating health effects of ^{90}Sr chronic ingestion cannot explain all of the health effects of chronic long term exposure in a post accidental situation. However, these data strongly suggest that this multiple exposure, including external irradiation at low dose rate and internal exposure with several RNs should be taken into account in future studies.

3 Radiological and chemical multiple exposures.

There is almost no available study about the health effects of a combination of RNs with either a chronic or an acute exposure. In fact, the ICRP consider the additivity of radiation doses, whatever their origin. This principle is in accordance with the direct proportionality between the frequency of stochastic effect and the radiation exposure. However, this principle does not take into account several points. The first one is the chemical toxicity of stable nuclides, which may become preponderant at low concentrations, as it is the case for Uranium. The second point is the description of non-linear effects at low doses such as bystander effects, hormesis or adaptative effects. And the third point is the existence of other chemical pollutants in the environment that may interfere with ionizing radiation effects, especially at low doses. For instance, it is well known that radon exposure exacerbate the risk of lung cancers in smokers as compared to non-smokers. As well a recent study showed an increased risk of ^{131}I-induced thyroid cancer if there is a co-exposure to excess of nitrates.

4 Conclusions.

The hypothesis of a more generalized model of interactions between chemical pollutants and radiological pollutants is reinforced by the recent descriptions of non-additive (either supra- or infra-) biological effects of mixtures of chemical pollutants. For instance, a study of 7 different combinations of 2-6 chemical pollutants showed in some cases supra additive effects that were not predicted on the basis of toxicological reference values. Since there is no available study on the health effects of mixtures of radionuclides, it is not possible to exclude the hypothesis that a mixture of RNs may induce non-linear, and possibly supra additive effects as compared to the effect of individual RNs. Thus it appears that futures studies on the health effects of radionuclides at low concentrations should be done taking into account more realistic schemes of exposure, including several radionuclides, but also mixtures of RNs and chemical pollutants. Such mixtures may explain some of the currently observed health effects of exposure in post accidental situations.

https://doi.org/10.1051/bioconf/20191408005

More effective use = more lives saved: medical countermeasure response strategies for mass casualty radiological incidents.

Juanita Jones[1], and *Chad M.* Hrdina[1]

[1]Assistant Secretary for Preparedness and Response, United States Department of Health and Human Services, 200 Independent Avenue SW, Washington, DC 20001, United States of America

Topic: medical countermeasures and decorporation

1 Introduction

Strategies for the use of medical countermeasures are vital to ensure maximum lifesaving during an incident response. Federal budgets for development and acquisition of medical countermeasures (MCMs) to chemical, biologic, radiological and nuclear (CBRN) threats are often inadequate to fund infinite sustainment of national stockpiles. In addition, necessary infrastructure, diagnostic capabilities, operational, and personnel resources can also be limited or unavailable due to lack of technology or inadequate funds for investment. Thus, federal governments and community emergency medical response and public health partners must develop strategies to leverage scarce MCMs and other resources to ensure a mass-casualty response approach to a CBRN incident that efficiently and effectively saves the most lives. The decision framework approach and MCM response strategies for radiological incidents will be presented.

2 Strategic Framework

Medical countermeasure response strategies are divided into three major portions with supporting information: 1) Notification, and 2) Scope of incident, lead to 3) an MCM response strategy, which is supported with MCM specific information, special considerations for special populations (e.g., children), regulatory or legal issues or needs for MCM utilization, as well as useful public messages recommendations, and current clinical guidelines summary.

2.1 Notification and incident scope

How we become aware of an incident is an indicator as to the potential affected population and severity of injuries or illness at the time of notification, which means we can anticipate the spectrum of casualties resulting from the incident, and make decisions for allocation for MCMs such as Prussian blue or Ca/Zn-DTPA. The nature of an incident, including factors such as time, affected population, ability to identify affected population, mode and magnitude of radiological dispersal are important because they allow responders and decision makers to identify and characterize the population that has been exposed, or at risk of exposure. This informs decisions to protect a population at risk, from any or further exposure, and to implement public health or medical interventions to mitigate or treat injury from radiation exposure or internal contamination with radionuclides. A very important aspect of characterizing the magnitude of an incident translates into an estimation as to whether enough resources are available or there is potential for a shortage. Scoping the incident is not about identifying all of the potential scenarios, but rather those factors that differentiate scenarios, such that one would change the strategic approach to response.

2.2 MCM response strategy

In the initial response period, much situational information many not be known, but decisions will be required to ensure assets, including medical countermeasures are quickly employed early. As more information becomes available, flexible decision makers can adjust the medical countermeasure response approach. The MCM response strategy discusses target populations and intended population-based use of medical countermeasures. Deployment and distribution of medical countermeasures may include strategies for prophylaxis, mitigation, treatment, of target populations with different medical countermeasures at different times. The MCM response strategy recommendations ensure efficient and effective use of medical countermeasures based on confirmed or estimated factors of the incident and address both adequate and scarce resource strategies.

3 Radiological incident MCM response strategy

Potential for scarce resources following a radiological incident is an important issue that requires consideration for MCM distribution strategies. We recommend that in a scarce resources situation, early decorporation remains an important mitigation step that should not be delayed for confirmatory bioassay to initiate therapy as a means to conserve MCM resources. Instead, our analyses support an approach where epidemiology and simple screening can inform pre-emptive administration of MCMs to casualties with suspected internal contamination, which can save more lives than bioassay confirmation as the initiation threshold for decorporation. We recommend bioassay for discontinuation of therapy, and there may be dosing alternatives, as well as alternative thresholds for discontinuation of therapy, to conserve medical countermeasures. Furthermore, conservation of resources means establishing appropriate thresholds for target populations, which could mean palliative care for those too sick to benefit from decorporation.

Likewise, response to an incident with sufficient resources availability for the internally contaminated population should also leverage bioassay for discontinuation of therapy where thresholds might be set far lower, and screening may include those with less risk. Regardless of resource levels, bioassay is should be prioritized in cases where confirmation is preferred before initiation of pre-emptive therapy, to reduce high risk for adverse effects of unnecessary MCM administration (e.g. neonates).

4 Conclusion

Deliberative planning the strategic use of medical countermeasures for a variety of radiological risks means that if a radiological incident were to occur, we are prepared to use medical countermeasures in the most efficient ways, with the most effective approaches, such that we maximize the lives-saved. Planning prevents lengthy contemplation during response that could impede timely access to medical countermeasures by those who need them.

AUTHOR INDEX

AUTHOR INDEX

AUTHOR INDEX